the Artist:
Faith, Science, and the Rest of Us

Matt Loveland

L///P

the Artist: Faith, Science, and the Rest of Us

Cover art by Kate Waddell. www.katewaddellart.com

First published in the United States by *The Loveland Press LLC*, 2020.

All rights reserved. In accordance with the U.S. Copyright Act of 1976, the scanning, uploading, and electronic sharing of any part of this book without the permission of the publisher constitute unlawful piracy and theft of the author's intellectual property. If you would like to use material from the book (other than short excerpts for quotations), prior written permission must be obtained by contacting the publisher.

Copyright © 2020 Matthew Loveland

ISBN: 978-0-578-76316-3

CONTENTS

1	The Rat and the Cat: A Love Story	1
2	Evolution 101	11
3	Of Boats and Kangaroos	26
4	Mars to Kansas, With Love	46
5	Fish Under the Ice	52
6	The Chicken and the Egg	57
7	Guilty by Association	67
8	Killed by Clothes	73
9	Atheism: The World's Most Boring Religion	86
10	Cover and Move	100
11	The Burden of Proof	116

12	The P Word	135
13	Stalin, Hitler, Mao, Pol Pot	147
14	For Which We Have No Answer	158
15	*The Creation of Harris*	167
16	The Carpenter Who Builds a Bridge	181
17	The Closer You Go the More You Find	197
	Acknowledgements	209
	Annotated Bibliography	214
	Connect with Matt	248

Chapter 1

The Rat and the Cat: A Love Story

Single cell parasites do not strategize, they can only react to stimulus in the environment.

In a back alley not too far from your house lives a little Rat named Scoots. Never mind how he got that name, just that he is in fact named Scoots. Our four-legged friend is not one of those brutish grey rats who likes to chew through the wires for your Wi-Fi or jump out and spook you when you were unsuspecting. Scoots is a tan rat with a cream underbelly, he lives in a hole dug under an old soup can and uses the lid as a door. Scoots is the friendly sort that, if he were to cross your path, would stop on his hind legs and give you a little turn of the head, a kindly ear scratch, a wink and a nod if you will. He is widely respected as a no-nonsense rat that can mean business when necessary, but always has time to stop and smell the roses. A real upstanding rodent.

One day, Scoots was about his morning routine when he came across a small bowl of tuna, left out for a cat named Muffin. He knew Muffin to be a mean, old cat for two reasons: one, that it was a tough male unfortunately named Muffin, and two, more unfortunately, that he was owned by the kind of girl that would name a cat "Muffin". The other rats typically made fun of that poor creature, but Scoots could find it in his little pumpkin-seed-sized heart to feel bad for Muffin and his lot in life. But nevertheless, he didn't feel so bad as to be above eating from the cat's bowl. Our hero scurried under the chain links at the bottom of the fence and snaked through tall grass at the edge of the hedge left uncut by the girl's loser boyfriend who unironically vapes while driving a moped.

The grass bent around his strong whiskers and he scurried up the wooden side of the steps, onto the porch and began to enjoy the rewards of his bravery. Self-satisfied, he crawled away into the shrubs and through the neighbor's garden. Before returning to his soup can, he knew that when he got close to the base of the billboard that awkwardly advertised male "health" to all the annoyed commuters languishing in traffic, he would have to make a wide berth around the abandoned cottage. A sweet old lady put out food for the racoons and cats and they all somehow agreed to play nice. A marvelous achievement in peace really. A treaty that did not include Scoots.

The next day our hero was talking with some friends, Jonathan and Carlos, under the overpass. Scoots had been feeling a little lonely recently, seeing the other rats he graduated with having gotten lady rats to settle down with and make children. When suddenly, along came Debby, that attractive rat who always found provocative—but still somehow awkward ways—to make her intentions known to Scoots. He met her gaze and she gave him the eye roll with some shoulder action. His intestines did the Macarena and he begged the pardon of his fellow rats and followed Debby across the sidewalk and onto the old railroad.

They spent the day doing lots of rat things and generally enjoying the grimy underbelly of society. They chewed through the internet connections of a household of loud hipsters and whispered truths about life to homeless guys who tragically would never be believed. All along Scoots could "feel it coming in the air" as we might say. The eye rolls, the flirty tail flicks, the shoulders. The shoulders!

Scoots would be afflicted by a dry spell no more!

As the night settled in and they dined finely over a packet of saltines and the remains of the beer drank by pansies, Debby made fierce eye contact. She suggestively pawed the ground, she scandalously rolled one shoulder and then the other. And all of a sudden.

Scoots wasn't feeling it.

Like a light switch, she was no longer the fixture of his fire. He made some shuffling excuse about leaving the lights on in his den and scurried away embarrassed. A hard day for our friend Scoots, by all accounts.

…But then the smell hit him. A certain kind of *je ne sais quoi*. An aroma that made wild his beating heart. Something was changing in him. Something was *alive* in him. Like some kind of creature was working the levers of his brain, some cupid, bent on true love. He put his great sniffer into the air and called forth the awesome and terrible power of his nostrils. His paws tore the ground on the hunt for his lover, for the object of his carnal passions. He was aroused like never before.

The billboard for male vitality was lit as if to welcome Scoots to the layer of his love. Something echoed in between our hero's ears but he couldn't put a claw on it. Never mind the doubts, Scoots was getting lucky. Except that Scoots would not get lucky.

Several very cute kittens, mostly orange and white except for one that endearingly was all white with a black nose, murdered him.

Scoots died because of a single-cell parasite called *Toxoplasma Gondii* (Pronounced: Toxo Plasma

Gandhi). Over the course of time it came to inhabit his brain. It then rewired our friend into being aroused by the scent of cats. This led Scoots to pursue a romance with a creature that wanted to murder him. It doesn't just remind me of my dating history, it also gets crazier than that.

The parasite does all this because it can only reproduce *inside of a cat.*

Toxoplasma Gondii hijacks the brain of a rodent, makes it attracted to cats, and gets the rat murdered so it can reproduce inside the small intestine of a cat. It cannot sexually reproduce otherwise.

Ah, the Rat and the Cat. A love story 10,000 years old. And also the signature of our Intelligent Designer.

You were created. Of that much I, nearly the entirety of human history, and the majority of scientists would agree. In fact, in 2009, the last landmark study by the Pew Research Center (widely considered the gold standard of non-biased statistical research and meta-data analysis) polling the prevalence of faith among scientists, found that 59% of all scientists polled from a wide range of fields believed in a "higher power" or a Creator. In 2018-2019 Pew conducted another survey asking people to self-identify their belief system. A whopping 4% claimed to be Atheist. If you were wondering why I care enough to write a book let me share one more statistic with you:

100% of all American Public-School students are taught—both directly and indirectly—to believe conclusions which are of enormous spiritual and scientific ramification that *not even field-leading Atheist scientists believe.* Jim Peebles, winner of the 2019 Nobel Prize for his work in the field of Cosmology, among his other achievements, developed the theoretical 14-billion-year framework of the universe. He was quoted at his reception, when asked about the Big Bang as an explanation for the creation of the known universe, saying, "It's very unfortunate that one thinks of the beginning whereas in fact, we have no good theory of such a thing as the beginning." Nearly every American child grows up in a world told that one of the most obvious pieces of objective truth, that we were made on purpose, is false. They are raised in the church of *Nihilism*, the belief that there exists no meaning in life, no higher purpose, no aspect of divinity that makes each human individual both necessary and sacred, in a word, *Holy*.

And by the way, while we're introducing ourselves, I'm not a scientist.

I decided to write this book after realizing that many, many young people were struggling with the same, big questions—and were getting the same bad answers and half-replies in return. Having spent years reading, studying, and picking apart the "Science" from what I came to see was really "guesswork," I felt pretty comfortable tackling most of these questions with the young people who were asking them....But, I knew that I would never be able to share this information with more than the few students I would be lucky enough to work with.

On a personal level, I came to see why young people are so confused about life's most basic questions—and why they are experiencing so much angst. They grew up in a world that is constantly telling them that their lives are nothing but random, cosmic, chemical "accidents"—and that the endorphin rush of "having fun" is the only, true meaning in life. I knew the effect this kind of nihilistic "teaching" had on young people. I often dealt with the fall-out.

My background is in the military, coaching, and youth ministry. My time in the service provided me with the experience of teaching complicated ideas, in easy-to-understand formats, to diverse audiences. Likewise, my years spent coaching athletics have given me countless opportunities to teach, explain, lead—and encourage—a diverse cast of young men, many of whom were (and are) struggling to find their "place" in an increasingly chaotic world.

Finally, I decided to tackle this subject because I am *uniquely qualified* for the task; put simply, explaining the complex and leading others *is* my background. And, especially these days, I believe that we all need reminders that our lives *matter*. That we were made *on purpose*—and that we're here for a *reason*....I will show you that, put together, there is only one, *logical* place that careful study of Science takes us...to our Creator.

So, how does the death of Scoots disprove the most self-destructive belief in human history? Because to believe it happened by chance mutation, selected for by the environment, you would have to also think that a parasite would randomly evolve, in the small-intestine of a cat, the super power of multi-stage mind control over a totally different animal than its own host. And remember that it's a single cell organism with no computing power or rational choice. And also remember that if the specific cell that just so happened to develop this miracle two-stage mutation found itself eaten by an earthworm, ant, bird, human, or nearly any other creature or experienced any other natural end that caused the cell to die without making it back to a cat, then the nearly impossible series of mutations would be gone forever. This is, as we say, impossible.

Chapter 2

Evolution 101

Evolution has nothing to do with religion, it is simply the best working theory for how you got to be reading this sentence.

I was always an average student. Especially in the sciences. I think it had more to do with having to memorize facts, and then spit them back out on a page, which I would no doubt have to relearn when I chose to take door number three in life and write a book about evolution. For those folks out there who like me were keenly aware of the fact that people didn't evolve to spend their adolescence sitting in school for eight hours a day, let's all get on the same page before we start taking swings at the scientific explanation for our birth into the universe.

Everybody picture a field out there. Nowhere in particular, just over-yonder-that-away. In that field live 10 field mice. And let's say that with every generation of field mice a particularly mean Owl flies by who will murder two mice and be a real jackass about it.

Now it stands to reason that the likelihood is high that this Owl will catch the slowest of the mice. In this generation a mouse is born named Sneaker. Sneaker's genes were mutated randomly and he is 1% faster as a result. Let this speedy gene be called SS. Whereas the normal gene is called ss. Before we get any farther, you may not know it, but for the Atheist that just read the third sentence before this one they might well be deeply offended by my use of the word "random" and with good reason. In the great debate over religions and school science classes, Christians, well intentioned of course, flippantly call evolution "random," in order to make the theory seem silly. Allow me to explain. It is random in one sense and not random in another. *It is* a random mutation that Sneaker was born being faster as opposed to having an extra toe or a third eye. He could have had mutations to any one of these genes but did not by no control of his own. His cells didn't think, "it would be prudent to make the little guy faster". They have no say in the matter. It *is not* a random mutation in that the environment (predation, mate choosing, etc.) would likely not allow a Sneaker to be running about with a useless extra limb or plaid-patterned fur.

It stands to reason again that Sneaker, with his go-faster gene of SS, would be the least likely to get snatched by the Owl and thus, more likely to reproduce. This is referred to as Natural Selection. "Slowness" in the mice is being "selected against". The environment is "selecting for" the mutation in Sneaker that lets him go 1% faster.

So, after the Owl is done, there are four breeding pairs of mice. They each produce four offspring, for the sake of all us liberal arts majors, and this results in sixteen offspring total, all except four having the ss or normal gene. Four would be born with a mix of the genes, demonstrated as Ss. The owl has three kids in the meantime but one is vegan to the great shame of his species. Now before we go on just remember that one of the Owls is vegan. Not for any real reason, just because he told me to make sure you were reminded.

Now then, that leaves us with two murderous Owls after their old man kicks the can. They each eat two mice. Now in this generation four of the mice are .5% faster and thus more likely to reproduce. There are thus six breeding pairs after four mice bit the big one. That means that statistically there would be two mice with genes of ss, eight mice with genes of Ss, and two mice with genes of SS. Meaning that most mice would be .5% faster, and two mice would be 1% faster. We could continue this thought experiment but the idea is the same and I'm starting to get a headache.

Evolution, put simply, is the change of a genetic composition of a population over multiple generations.

Now if we go back to Sneaker's progeny, we see that many mice are very fast, whereas some mice are very slow but way more stealthy. In this example of *Disruptive Selection* the "normal" mice in the middle that are neither fast nor stealthy are being selected against and are quickly eaten or bred into one category or the other. The stealthy mice think the fast ones are rude and immoral, whereas the fast mice think the slow ones are lost in the past and not progressive enough. This creates two populations that choose not to breed with each other. After enough time, they will become two separate species.

This process began with chemical reactions in the ancient oceans and led to *Abiogenesis*-or the advent of living organisms around 4 billion years ago. Those single-cell ancestors mutated and grew in ways which made some more biologically fit than others. The unfit ones died off, the fit reproduced. These mutations led eventually to multi-celled organisms, which became more complex across the slow drumbeat of time. Mutations eventually became such that individuals and their communities became unlike others—to the point that they were their own species. Over time, this process has led a few bacteria to the populate the entire world with all the species therein. That, in a nutshell, is evolution.

Evolution says nothing about religion. It does not ascribe meaning or purpose to anything. Evolution is just the best theory out there for how you got to be reading this sentence. It says nothing about why you feel the way you do about the sentence, or how you should respond to this book. In short, evolution does not *do* anything. Evolution is not a person, place, or thing.

When I was a junior in high school, I watched some videos about how bad life was in West Africa, and specifically the dire straits some of the villages were in, following war and strife in the region. I had some advanced first aid under my belt, and so I decided to sign up on an outreach trip to go to Sierra Leone. My parents signed off on this for whatever reason, and that July, away I went.

I didn't sleep on the red eye flight to Belgium because the malaria medicine I was taking was having a negative reaction with me so I discontinued it (foreshadowing), and then we flew to Freetown, Sierra Leone, a city of over a million people without a sewer system or a power grid. That night I slept in a blanket that didn't cover my calves or feet. I had the sense to cover my face which paid off because I was bitten by mosquitos so much that when I woke up everything that wasn't covered by the blanket was swollen. We didn't eat nearly enough. The villagers, God bless them, shared what they had, but it wasn't much. We saved the lives of dozens of people who could not afford doctors and I have never been so widely appreciated by a group of people in my life.

A week after getting back, I started getting sick at night. It wasn't bad and the outreach director said I would be fine. It went on for a couple nights and I shrugged it off. Six months later, while playing a game of Rugby, I got pulled out because I was so dizzy. I sat on the bench and then started shaking uncontrollably. I was so cold I couldn't believe it. My teeth rattled in my head, colors were mixing, and I started going in and out of consciousness. A doctor on the sideline talked with me for awhile and said it was likely that I had gotten the Plasmodium Ovale strain of Malaria (their memories are a great wonder to me). The rarer of the four forms, it hibernates in the liver until it wakes up and breeds in your brain. It can manifest itself for years and so it did. It left me dizzy during the day and hallucinating and shaking at night, and I became symptomatic several times over the course of three years. Worse than the flu for sure but altogether better than my ex.

Toxoplasma Gondii is similar to Malaria, in that it's a parasite which uses the cells of its host to live and reproduce in. The cat is the only *definitive* host of the parasite—which means the creature can only sexually reproduce in the cells of the kitty's small intestine. This means that two *T. Gondii*s can make kids who will have both negative and positive mutations over time which will be naturally selected for and against, generally improving the species over time.

The way it gets into the kitty is that parasites form cysts when they mate and (I apologize if you're reading on your lunch break) the cat poops out the cysts. These microscopic eggs, if you will, get on the ground, plants, and pretty much everywhere by the wind and rain, and rodents inevitably get infected by them. At this point, the cysts mature and infect the *intermediate host*. In the rodent, this causes a strong sexual attraction to cats and it results in them being horribly murdered like Scoots. This is made possible by the parasite making two key changes to Scoot's Medial Amygdala (part of the brain). One, that Scoot's would be much less averse to scents like cat urine, and two, that he would develop a sexual attraction to cats. Most experts agree on the method of brain control, here scientifically called *epigenetic remodeling*.

In humans, dogs, fish, and nearly every other life form that can encounter these cysts, the parasite cannot reproduce sexually so it has to clone itself asexually. It can become symptomatic in some cases, and lay dormant in most others, and, specifically, is found in an estimated 30-50% of the world's human population. This asexual cloning has led to three very specific strains of the parasite, due to the identical copies it produces through the generations.

That is, in its most accessibly basic form, the parasite *Toxoplasma Gondii*. I have just laid out what it does with quality citations to known experts (all accessible in my *Annotated Bibliography*, in the back of the book). I can't cite what I'm about to say because it has no citation, because there is a total absence of knowledge in the subject. And that is *how T. Gondii evolved.* It is a complete mystery. Any scholarly paper you fight your way through will eventually say something to the effect of, "More studies are needed to produce models for how *T. Gondii* evolved." Or more satisfyingly they will make rough guesses as to its origins.

In my research, as I talked to different biologists, I was assured that there are many parasites that time their own reproduction cycles to other organisms. Of this, I have no doubt and having majored in History I make myself and scientific colleagues of mine laugh constantly with all the holes in my knowledge of what I'm sure are basic concepts. But there is an important difference between myself and 41% of the scientific community.

When made uncomfortable with something in my life or in the news, I've never concluded there was no God. When I was young, I just assumed He was bad. The entire New Atheist movement is built on theories that have no beginning. You have never heard of how the Big Bang started because there does not exist a theory for how it started. We can think about finches all day but it gets us no closer to describing a philosophical ship that can float us away from the island of God.

There are two smoking guns for an Intelligent Designer: the two ways that *T. Gondii* can control the brain of another organism that is not its definitive host. If the rat were simply less averse to *all* predators and biological threats, I might throw out the argument altogether, and if the rat developed sexual attraction to other animals *in general* this would make the parasite more biologically fit and would be way easier to conceptualize. The fact that not only are these attractions and lack of aversions specific to the cat *only*, and that both are represented in the infected rodent, it leads me to conclude that these evolutions are only possible in a gene sequence that was rigged from the start.

There are 6000 genes sequenced in the DNA of *T. Gondii*. Every time the parasite reproduces, one or many of those genes can mutate. But this only refers to the process of totally random mutations effecting miniscule changes that continue over millions of years to affect real change and create new species.

But these changes must be possible in the first place. Why didn't *T Gondii* simply make the rat more likely to clam up in the sight of something that made it afraid? Or just make the rat less fast? Both forms of mind control, the latter being a simple *governor* in electrical talk, would produce a Scoots less fit and a parasite more likely to end up in the belly of a cat. But instead, it is a two-factor brain control which *specifically* targets one animal. And the specificity of that attraction is the end of non-theistic evolution as we know it.

"The parasitic strategy of T. gondii involves securing a permanent residence in the host and awaiting transmission." Reread that sentence. *Anthropomorphizing* is a common cognitive dissonance where the speaker attributes to a non-human actor human qualities. *T. Gondii* is a single-cell parasite. It does not *strategize* anything. It does not *wait* for anything. It has no concept of time at all. Yet this is without a doubt the most concise and well-written accounts I came across in my study, widely considered a landmark paper on the subject. So much so, that when it was recommended to me by the head of a Biology department at a prestigious university (I don't want to out anyone as a collaborator with an Intelligent Design book) I had already saved the scholarly article to my computer and cell phone, along with having already cited it in my bibliography.

In the research I have done in Biology on average the more tenuous the claim of the scientist, the more they attribute human qualities and motives to non-human actors. Its comical but the fact that scientists can't avoid using the language of strategy and design is further proof of my point. There was strategy, and there was design. In His infinite wisdom, God, our creator, rigged the game. He picked winners and losers. He balanced an eco-system which is still ticking four and a half billion years in. Everything that you know was influenced by Him.

Chapter 3

Of Boats and Kangaroos

The Bible can be true without describing literal fact: God continues to enlighten the world, both spiritually and scientifically.

I grew up in a Southern town and around a group of people that ranged in between the Old South and the Modern Midwest. There was a nuclear plant near where I lived so intelligent people from all over the country and world moved to town for high-paying jobs in the field. The result was a strange mix of new ideas and old, particularly when it came to politics, football, and faith.

If you ever find yourself in Aiken, South Carolina, I can sincerely promote two things: One, the annual Aiken Steeplechase—a series of colorful horse races where fun can be had with a group and your own pick-up truck for a tail gate. Two, at the local microbrewery, aptly named, The Brew Pub. The Steeplechase is an important event every year that, and I mean this when I say it, draws at least half of the entire town to its sun-soaked fields. The single-lane roads of the sleepy, Southern town are engorged by traffic as thousands of pick-up trucks descend on fields used only for the purpose of that one race (to my knowledge, like most open areas in the South, hay is grown and harvested after the races, which leads me to suspect the race is planned for the beginning of the growing season).

Nearly every social organization in the city is represented at the races, including many churches and youth groups, looking to give the teetotalers a respite from the near-savage levels of alcohol being consumed. Having spent my formational years riding horses and working their stables, the event was to me like the Super Bowl. Long before I could legally drink, I kept the books for the gentleman's wagers on each race.

Neighbors, family friends, deacons of churches, teachers, and all manner of associations would wander through our impressive tailgate spread and share drinks and stories. The event was a melting pot for all the Northern, Western, and foreign expatriates who had come to town for a promising career in Science and Engineering.

The second reason to visit Aiken is a small bar tucked away in the quaint, but ever-busy, downtown. The Brew Pub is an old brick building built along the main drag with a pedestrians-only alley of cobblestones on its side. They serve a rotating tap of brews only available in-house. It's a popular drinking hole for people in town on business, workers from the Nuclear site, and the old-guard of the horse town with all the hired hands and riders alike.

And hanging above the main bar, nestled in between all the sports memorabilia and flat-screen TVs, is a Michigan State Spartans flag with associated miniatures and paraphernalia. The town is an amalgamation of different cultures, creeds, ethnicities, and backgrounds. Like much of the south, there is a church within a stone's throw of nearly any point in town, but those churches span the wide gambit of beliefs and sects. There is, of course, a broad spectrum of faith levels from fundamentalist to agnostic, and a sizeable portion of Atheists. The vocal minority for that part of the country is, of course, the Evangelical Christian community, and to say that teachers walked on eggshells in the classroom is something of an understatement. I say "eggshells" here because we were of course taught evolution along with classes on all world religions and the topic of Atheism was broached many times, especially in the sciences.

What separates that town from most is that obviously being the South there existed a much deeper pool of Christians that were much bolder than in, say, Seattle. The classes weren't restricted to teaching Intelligent Design or Creationism, like some schools in smaller Southern towns pock-marking the different stretches of the Bible belt. Even the private Christian and Catholic schools taught evolution, albeit with an obvious and understandable bias towards the professed faith of those institutions.

This all to say that I am something of an expert on being caught in the middle of this argument.

Creationism is the belief that the Judeo-Christian God (also referred to by some more specifically as *Young-Earth Creationism*) created the entire world in six, twenty-four-hour days, before He rested on the seventh. It takes the entirety of the Old Testament and New Testament to be literal truth. Their main arguments center on the impossibility of the creation of matter, and the great flood.

In the *Impossibility Proof* the Creationist would argue that due to the laws of thermodynamics, matter cannot be created or destroyed. Because, by nature of the existence of the universe, matter would have to have been created. Put another way, when I was in second Grade, I had a truly wonderful and kind teacher fresh out of college. He once talked to us about Creation when asked for his opinion. He replied that when he was in college he asked his science professor where they thought everything came from. The professor was the classic fundamentalist villain: a jaded, looks-down-their-nose-at-the-commoners socialite and Atheist preacher. He did not, I'm almost positive, provide those details but his description fit into the box we had been raised to create. The professor answered that everything occurred through the crash of rocks in space at such high speeds that it caused the "Big Bang" (this is a completely true story by the way. I don't know how I remember it either.) My teacher then asked, "Well if that's the case then where did those rocks come from?"

The Professor, "Well they came from smaller rocks of course."

My teacher, "And those rocks?"

Checkmate.

Or so it goes, I guess.

The idea is that something cannot come from nothing. If the rules we have do not fit the bill, then throw the rules out and write something that fits.

Growing up one of my favorite shows was the Discovery Channel classic, *MythBusters*. A fun cast that tested out popular myths you heard about in every day conversation hosted the non-fictional show. "After falling for 100 feet is hitting water like hitting concrete?" Nope. "Will throwing a penny off the Empire State Building (even if it could clear the levels of the lower floors) kill someone?" Also no, but every year the building clears all the thrown pennies off the lower levels and donates them to charity. Think about the last two sentences. I'm not sure how I feel.

In the show they had two basic goals: replicate the myth, and, failing a positive outcome, replicate the result. If a penny falling that far would not accelerate fast enough to kill someone, how fast would it have to go? They then would answer the second question for entertainment and science value. They can't satisfy the original conditions of the question, so they make a new question in which to fit their answer. The same is largely true for both the Creationist and the Atheist. The Creationist views the problem of matter and simply says, "God made everything and it is beyond our understanding". Assuming there is in fact a Creator, that explanation is possible but does not satisfy any higher scientific questions. The Atheist declares that there is an answer to how everything came from nothing, we just are not yet (or perhaps will never be) able to totally answer such questions, and they don't believe in a higher power for other, in my opinion, largely unscientific reasons that we will get to later on.

The Great Flood from Genesis is the idea that all of mankind was wiped out by a flood that covered the whole world. Interestingly, over 16 separate people groups have "Great Flood" myths that largely agree on the time and scale. Most historians conclude that there likely was a massive flood event, most likely caused by a breakdown in the ancient land barrier between the Mediterranean Sea and the freshwater basin of the Black Sea. The result was the cataclysmic destruction of the ancient human civilizations that farmed the region, and is consistent with the diaspora of Indo-European people groups across the Middle East around that time. Without written history, it's an obvious logical step from localized catastrophic flooding to God-decreed destruction of the whole earth. Creationists would take the first part as the grudging proof of the science community, throw out the smaller scale, and state that science puts the win in their column. Ken Ham's "Answers in Genesis Foundation" runs the Creation Museum in Kentucky and has a life-sized replica of Noah's Ark. In it, the Creationist would say that God literally took a breeding pair of every animal (they could be adolescent versions, saving space and food) and saved them in an ark that floated over a flooded world. They will tell you that there were on board the "parent species" of these animals (one or several ancestors of canine, as opposed to thousands of different breeds of dog and wolf, for example) and these animals all stayed together on that ark. Notably they also claim that dinosaurs, never mentioned in the Bible, were also aboard, and able to fit.

The ark has specific dimensions listed in the Bible and are as follows: 510 feet long, 85 feet wide, and 51 feet tall. There are one million known species of beetle. Let's assume you just took one animal of every species, including insects and flightless birds, never mind how penguins and kangaroos get to the ark (many Creationists believe in Pangea, or a super continent, just that it split apart in between the flood and now). In this scenario you would have to house all of them with enough food and water to survive the forty days and nights. You could say, I guess, that they were all in some kind of hibernation supported by God, but then we still have the space issue of the animal's mass. The second problem is that if the world was flooded by freshwater, all the animals of the sea would die from the lack of salt. If it were flooded by saltwater, all the freshwater fish would die.

For those reading who feel uncomfortable with this chapter I want to here extend an olive branch: all I am saying here is what we know and have studied. There may well come a day in the future where some revolution in science makes Young Earth Creationism a viable theory for how the world was made. Today it isn't. And I beg you to keep reading if this is a hard argument to swallow. We all agree that there is wisdom in listening to something you don't necessarily agree with. For those who do believe in a young earth, I would simply ask you to question your preconceived notions and explore what the majority of believers think. Obviously you don't, I assume, believe that all, most, or even a sizeable minority of scientists are actually evil and are maliciously falsifying documents and studies to dissuade the faithful from church. I'm just writing in this book what we know and what I think. Trust me, if I were just in it for the money this would be a book about Creationism and would be nothing but a 200 or so page pander to an audience with a robust infrastructure for marketing and selling content. My writing would certainly land better there. The church crowd is conditioned from a young age to laugh at dumb jokes.

My concern about Creationism is that for Christians it seems to be the default belief of fundamentalist and evangelical believers, not because it makes the most sense or even most accurately aligns with the text, but because the individual is applying literalism to what was intended to be understood of as a metaphor. It's not that they are "too deep" into their belief system, it's that my fellow believers have not yet gone quite deep enough into the text of Genesis.

Sometime around college I started looking online for arguments to reconcile problems I had with the Bible. And then I read and studied the ancient text of the Epic of Gilgamesh. It's a fascinating story that has all the good stuff: betrayals, romance, twist endings. It's a way for the ancient Mesopotamians to reconcile their own ideas about community and honor in a story that proliferated their entire culture and apparently many others. It also reminds me a lot of Genesis.

I think the book begins in a heavy metaphor and gradually turns into a literal account. Many men and women wiser than me would tell you that Genesis was inspired by God to tell a story that while true, doesn't happen to account real events the way they occurred. That person would likely cite something along the lines of: "Regarding human waste, God stated: 'A private place should be designated for use outside the camp, and there is where you should go. A peg should be part of your equipment. When you squat outside, you should dig a hole with it and then cover your excrement.' From Deuteronomy 23:12-13. God could not possibly have described microorganisms to a culture that didn't understand the most basic concepts of biology, so instead, He would just make a rule about being unclean. Moreover, that belief is consistent with the common Christian notion of God's continuing revelation to mankind. That Jesus in the New Testament would free people from many Levitical laws as well as their punishments, or how he declared that men were allowed to have multiple wives because of the "hardness of their hearts" but are no longer permitted to do so. The changes, to me, reflect the evolutionary nature of mankind biologically, intellectually, and, in my opinion most important, spiritually.

An honest, non-believing historian would likely agree that Genesis is a metaphor that gradually gains elements of historical fact before turning into a rough mix of history and myth. Much in the same way that parts of the Iliad are considered historical fact and are mixed with lots of creative license and religious overtones.

This, of course, presents us with a problem likely more important to the fundamentals of the largest religion on earth: namely, that if Genesis is a metaphor, than what of its most important story: the Fall of Man? Evolution, I think, allies closer with this doctrine than near any other, and quite possibly gives us the age of man, or at least "man" in any sense that matters, as being within the last 10,000 years. For several hundred thousand years we have fossil record of homo-sapiens using tools and eventually fire. Sometime in the last few thousand (indeed it quite possibly was the time creationists and biblical historians suspect—6,500BC roughly) I believe God so influenced the lesser homo-sapiens and our direct ancestors to think in ways instead of "I must/I must not" to "I ought/ I ought not". If it was an evolutionary spike not just of intelligence, but of something *spiritual*, it in fact would have been just one ancestor, or two in the case of a God who had big plans in store. And lets call these two Adam and Eve while we're at it. The change would have been revolutionary. Suddenly you no longer have very intelligent chimps- (I know they weren't chimps, stay with me)- you have humans with a moral compass, you have people living *on purpose*. They easily would have stayed alive and rocketed their way up the stone age food chain.

Suddenly, a few verses that always bothered me, Genesis 4:14-17, begin to feel newly justified:

> "'Behold, you have driven me (Cain is talking, he's the one who killed his brother Abel, they were both sons of the first man, Adam) today away from the ground, and from your face I shall be hidden. I shall be a fugitive and a wanderer on the earth, and whoever finds me will kill me.' 15 Then the Lord said to him, "Not so! If anyone kills Cain, vengeance shall be taken on him sevenfold." And the Lord put a mark on Cain, lest any who found him should attack him. 16 Then Cain went away from the presence of the Lord and settled in the land of Nod, East of Eden. 17 Cain knew his wife, and she

conceived and bore Enoch. When he built a city, he called the name of the city after the name of his son, Enoch."

A couple questions pop up. If Adam and Eve have only had two sons (that we know of, but heck for the point of an argument lets say there are two dozen), then from whom will Cain be a fugitive? And how does the land of Nod already have a name? And where did his wife come from? And who populates this city he builds? Either the city is a dramatic over-exaggeration, or there were enough inhabitants to come and inhabit it.

To all these arguments one might say, "Well God describes the first men made, but certainly never said he stopped making them afterward." I suppose that's true but it seems (at least to a dope like me) to be fairly noteworthy that at least one sentence would be dedicated to where exactly all the hapless civilians came from. And its not like the Bible does not contain information which at times seems tedious (looking at you "instructions on building the Tabernacle"). I think evolution gives us a more satisfying answer: men, lower homo-sapiens that they were, possessed already the rudiments of what would become, and what is now, human tribal behavior and communication. They may not have had moral purpose, but they certainly could communicate objectives and would know to stick together. Cain would rightly fear being a loner in a tribal world of creatures with stone weapons that thought as far ahead as the next winter, and the backing of a merciful God would ensure his genes get passed on.

Cain could take a wife of any of these savages (with full consent I have no doubt, this extra-intelligent man with a good stone-swinging arm I'm sure would have made a sought-after mate) and the children (who doubtless would possess the dominant trait of having a brain capable of thinking in complex moral and spiritual dimensions) would be "raised" as it were, with their father. These early men would, raised or not, doubtlessly prefer the permanent structures of a city named Enoch to the sleep-under-the-stars-get-eaten-by-tigers wildlands. Evolution, remember, is only a working theory on how we got to be reading books and giving them one-star-hate-reviews on Amazon. It says nothing of religion.

Here, I suppose the fundamentalist would say, "Well, how convenient for you to go snipping the Bible here and there. What is to stop you from doing it to things you just don't like." The answer is, of course, nothing, Christians are almost better at doing that then they are at having bad taste in music. But figuratively, and more to the point, I do not find myself "snipping" anything from the Bible. To say that the earth was created not in six literal, 24-hour-days is to interpret what I assume is a metaphor in a way that aligns with the fact, that, it is indeed a metaphor. When Jesus said to forgive your neighbor not seven times but seventy times seven, believers everywhere refrained from putting a tally down for every sin they forgave until a total of 490 transgressions had been overlooked.

As I've said, I was once a creationist and a favorite argument of mine was, "The Bible is a house of cards: take one out and the whole thing falls." Not the best metaphor when describing the oldest complete document in human history, one that has outlasted—with its adjoining people group intact—the likes of the Babylonian Diaspora and Hitler's Final Solution. But the idea stands to reason that if you do not like one chapter in Genesis and try to explain it away, you might leave out some uncomfortable passages against sex outside of marriage or paying taxes. And that idea is sound enough, except that it doesn't work in the context. The entire Old Testament tells the world the story of how man fell from God and continued to run away from Him. The New Testament is the story of how God fulfilled his promise to Abraham to destroy Himself if Abraham and his descendants didn't fulfill *their* promise to God. To write that story, God had to tell the beginning. The people who first heard these tales by firelight in the desert were ones that had neither Algebra nor germ theory. They knew no more about the stars and planets than what they could observe on a clear night. I'm not saying it would have been challenging to teach these people about universes, ten decimal places, or evolution; I'm saying it would be impossible. "But, 'All things are possible with God'" you might say. Well first that is a quote out of context, and second, as C.S. Lewis would say, "Nonsense remains nonsense even when we talk it about God." Impossible? No I suppose the Bible could have theoretically been a collection of books on the working of the universe and time and germs and politics. But graciously God made it a story about His love for the world and our response to Him and others. I believe we will find that being the only thing that matters in our time or any time.

Chapter 4

Mars to Kansas, With Love

Two things are true about time: it must have a beginning, and it can never go backward.

Before we move on to anything else, we're going to have to get into what even is *time*.

Sean Carroll, a Physicist from Cal Tech, spent several years trying to pin down the slippery topic of time—something we all experience but have so much trouble understanding. In many ways, the topic feels like trying to imagine yourself doing a chore or talking to someone but from the perspective of watching yourself do it. In his interview, he gives a summary of his work on defining time,

We remember the past but we don't remember the future. There are irreversible processes. There are things that happen, like you turn an egg into an omelet, but you can't turn an omelet into an egg.

And we sort of understand that halfway. The arrow of time is based on ideas that go back to Ludwig Boltzmann, an Austrian physicist in the 1870s. He figured out this thing called entropy. Entropy is just a measure of how disorderly things are. And it tends to grow. That's the second law of thermodynamics: Entropy goes up with time, things become more disorderly. So, if you neatly stack papers on your desk, and you walk away, you're not surprised they turn into a mess. You'd be very surprised if a mess turned into neatly stacked papers. That's entropy and the arrow of time. Entropy goes up as it becomes messier.

Go ahead and tuck the second law of thermodynamics away in your brain's back pocket (or purse if your brain wore a skirt today) because we will be coming back to it later. Its hard to over emphasize how important it is to remember that time is in fact an arrow going forward, we're defined by our inability to go backwards. But that does not mean that time goes forward *exactly the same way* for everyone, everywhere.

If you were to fling yourself into the heavens and colonize Mars, we imagine that in between all that homesteading and social media posting that you would get a bit nostalgic. At that point you might say to yourself, "I wonder what ol' Bob is doing back in Kansas *right now*" Except that the phrase *right now* doesn't actually mean anything in any real sense of the word. That is because you are experiencing time on Mars differently than Bob is in Kansas. You will have aged more than him, though your experience of time would be normal. This is a larger example of a principal that is true here on earth that we can prove. An atomic clock put into outer space came back a little bit behind the same one on earth. The cause of this is called *Gravitational Time Dilation,* which means that the gravity of an object affects the passage of time—so that with more gravity, you get slower time. If you do not have a head ache yet, then hang on for another more abstract illustration: imagine the "fabric" of time is a towel, one of those big fluffy beach towels, laid out on flat sand along the beach. Now take a water bottle and drop it in the middle of your towel. The water bottle is acting like Earth or any other large mass in the Universe. The towel will bend along the object and if you were to draw a line across the sand underneath the towel that line would get stretched by the object now bending the fabric of the towel. If a grain of sand were to move along that line somehow between the towel and the sand, its path would have become longer because of the bending of the water bottle; and that is as good an illustration of the bending effect of gravity on time as I'm intelligent enough to produce. It does not matter if you totally understand that illustration, if you don't than take my word for it that there is a distortion on time which occurs around gravity. You should here take my word for it because I have one of those trustworthy faces, but if you choose not to, I recommend Carlo Rovelli and his book, *The Order of Time*. And both my explanation and illustration are obviously a wild over-simplification, but they advance our intellectual football, so that I can say something more: time is relative to every cell in your body. To say you are experiencing time differently than me is really to say the average of your cells is experiencing time differently than the average of my cells. To say *right now* for humankind is only to use a common phrase that allows us to speak in averages.

One more idea before we move on to the reason I bring any of this up: nearly the only thing we truly know about time—other than that gravity affects its passing—is that it travels like an arrow. It can't reverse course. It must have begun and is still going outward. It is imaginable that one day it may well end if all matter truly tends towards disorder. But it remains that time must have begun. Sean Carroll will tell you that our

Universe came from somewhere else: that the Big Bang came from a different universe and created everything we now know and observe and interact with. This solves the immediate and obvious problem of what started the Big Bang and how the material got *here*. It does not, of course, serve any useful purpose because it does nothing to solve the now bigger question of *where did* that *universe come from?* We know that matter cannot be created or destroyed. So the "multi-verse" theory is all well and good if you seek to say that this was not the first universe, but to cite it is only to create more questions than you answer, which, as it happens, is something of a trend we will get into later on down the line.

Chapter 5

Fish Under the Ice

Whatever built or started the four dimensions we are constrained by, must exist outside of them.

So, as long as we're willy-nilly tackling deeply theoretical problems in the realm of space-time, let me introduce my favorite: the Creator of the Universe had to exist before His creation did. Before time, itself. I do not doubt that before anything was, He is. But God was certainly not the only thing that, if not totally inherent, existed before time. And I call that something math.

When I was five I took an Oreo at snack time (back when snack time was cool enough to hand out disks of simple sugars and fats) and split it in half.

"There, now I have more." I half-heartedly exclaimed. I knew something was fishy about my hip-fired calculations, but I was a scientist on the cutting edge, risks had to be taken. Immediately a classmate said, "No you didn't! You have the same amount, just now in two pieces." It's a constant in my life that whenever I say something silly, there is always a know-it-all on hand to put me in my place. Perhaps for the best. But, regardless, I did my calculations, and realized that this boy, likely an IRS agent or DMV worker now, was correct. Math applies whether you like it or not. It does not care at all if you happen to be 5 or 55, a small child or a boulder or a pelican or an over-priced ice box. A drop of water must always fall from the clouds, it doesn't get a vote in the matter. The fourth minute will always follow the third. If there was matter before time and *entropy* (the decay of one ordered thing into a more disordered one), then two separate bits of that matter would be considered two total.

1+1 will equal two in any dimension. Another way to think about this is that the word "dimension" here is literally the dimensions that you measure for something.

I will review what we *know* of space and time and add to it what the consensus of scientists *are led to believe*. In the first dimension exists the line. We can only represent matter by its length along an infinitely narrow line or lines. In the second, you get an additional factor: *width*. Here, you can draw shapes like a circle. In the third, you get the addition of *depth* which allows for the drawing of *objects* like the sphere. If you were asking for the dimensions of a box, you would be asking for its measurements in the third dimension. In the fourth dimension everything you know of is able to be represented because of the new measure of *time*. In the fourth dimension my hands can type these words and your eyes can follow the sentence to read them. It is speculated and demonstrable that there exist more, and while it is a somewhat controversial and messy topic, I will leave this primer, seeing as though four, with the possibility of another, is enough for our purposes (and even at that we're beginning to leave the realm of what I'm passably intelligent enough to speak on).

If I were to take two separate lines of equal length and combine them, I would get a line double the length of one of the lines I used to make it. That is 1+1=2 in the first dimension. I can replicate that in the second, third, fourth, and so on as long as we have dimensions in which to "so on".

But, time only exists as we know it, beginning in the fourth dimension. We then realize that math is in places that time is not. Math is part of the fabric of reality in the most real sense imaginable. In this way, I say that math is beyond our ability to comprehend as we are expressed in the fourth dimension and can't conceptualize what the fifth could be. Let me make an analogy: if you were a fish that was born, lived, and died in an ever-frozen fish pond, then you would never conceptualize the vastness of the world that lay beyond, and all the different types of being that go on there. This is all to say that we're the fish in that pond. We, in the most real sense, are the math of God and are expressed in this dimension. We can't imagine the dimension of our creator any more than He can be confined to our world. The boy who draws a triangle could never exist in his conscious form on a bit of paper. God is no more "part" of our universe, than is the painter "part" of his easel.

Chapter 6

The Chicken and the Egg

The catalyst for the Big Bang must have existed before the actual event.

So, in review, we know that time can only go forward and must have started. We know that math exists where (and if you will excuse me for saying it, when) time does not. We now can move into the meat of the matter. One thought experiment comes close to stumping both the Intelligent Design scientist and the Atheist: what came first, the chicken or the egg? Or, put another way, who created the creator? Whether that creator is God or the infinitely dense point from which the whole universe is said to have exploded, you have two vying creators and both of them make claims to having pre-dated that which is.

Nothing you have read in the last two chapters is deeply controversial to anyone in the know. My next point is: what, or whoever, started the Big Bang had to have started time, and to have existed before it did.

To the Christian this is probably an obvious statement, to the Atheist its likely not but to both parties, I would ask that we remain on the same page. The truth of the matter is that there is not a "working theory" for how the Big Bang started. There are only ideas. And most quotes I have read go like this: "All matter existed in a singularity in a point smaller than a basketball and then the Big Bang happened and created everything we know of as the universe rapidly expanded outward and is still expanding." And before the claim of *strawman* "An intentionally misrepresented proposition that is set up because it is easier to defeat than an opponent's real argument," according to www.dictionary.com, gets thrown around, I'd like to provide one of the most concise explanations I have found for the big bang.

Dr. Paul Sutter, who according to space.com "is an astrophysicist at SUNY Stony Brook and the Flatiron Institute in New York City. Paul received his PhD in Physics from the University of Illinois at Urbana-Champaign in 2011, and spent three years at the Paris Institute of Astrophysics." Writing for an article on the big bang in 2015, says:

> The Big Bang theory can be summarized thusly: At one time, the entire universe — everything you know and love, everything on the Earth and in the heavens — was crushed into a trillion-Kelvin ball about the size of a peach. Or apple. Or small grapefruit. Really, the fruit doesn't matter here, OK?
>
> That statement sounds absolutely ridiculous, and if you said it a few hundred years ago, well, I hope you like

barbecues, because you're about to be burned at the stake. But as crazy as this concept sounds, we can actually understand this epoch with our knowledge of high-energy physics. We can model the physics of the universe at this early stage and figure out the latter-day observational consequences. We can make predictions. We can do science.

At the "Peach Epoch," the universe was only a tiny fraction of a second old. In fact, it was even tinier than a tiny fraction — 10^{-36} seconds old, or thereabouts. From there on out, we have a roughly decent picture of how the universe works. Some questions are still open, of course, but in general, we have at least a vague understanding.

One man's *question* is another man's *logical fallacy*, I suppose. The idea is that all matter was condensed to a ball and became unstable. According to Stephen Hawking, another possible theory is

> There was never a Big Bang that produced something from nothing. It just seemed that way from mankind's point of perspective... ...Events before the Big Bang are simply not defined, because there's no way one could measure what happened at them. Since events before the Big Bang have no observational consequences, one may as well cut them out of the theory, and say that time began at the Big Bang.

And the late Dr. Hawking certainly had a point: the only way the Big Bang works is if you are also willing to accept that the matter of the universe is *inherent*. Not just in the Alpha and Omega sense of the Judeo-Christian creator, but in the sense that unique particles and atoms existed in at least a precursor form from *always*. That the "peach" ball of matter simply *was*. We see time as an arrow shooting ever outward but it must have originated at a basic point. As far as we can tell, it moves linearly. We know only that a person can't travel backwards in time, so we arrive again at square one.

The problem is that of basic physics. For an object to move initial energy must have transferred into that object in order to move it. For that point to then explode outward, it must have received energy from somewhere else. If a boulder rolls down a mountain, you might well say that, having grown up skiing on those slopes, that the *boulder was always there* (but what you mean is that *the boulder has been there a long time*). The boulder released its *potential energy* (that it had the capability to move to the bottom of the hill) by rolling down-slope. But unless placed there by something or someone (a particularly large glacier or a powerful helicopter in the case of a smaller boulder), it ostensibly got to the peak by the slow, tectonic movements of the earth, and rode the land up until it formed a mountain—and was only dislodged by the gentle and nearly imperceptible changes of the geography of that mountain. But this is still energy moving the boulder, albeit so slowly as to barely warrant mention. Therefore, it was over the ages that the boulder was carried up into the heavens and then received its catalyst (a tiny shift in the plates, erosion by the elements, foreign tourists taking dangerous selfies) and released its potential energy to move downhill and smash through the delivery truck carrying that new pair of shoes you were waiting for.

A favorite thought experiment of Intelligent Design proponents is the "Junkyard Tornado" which paradoxically was the invention of Atheist Fred Hoyle (he hilariously used it to argue not for creation or evolution, but for ancient aliens placing life on earth). The idea was quickly adopted by the old and young earth thinkers and it goes like this: a tornado sweeping through a junkyard with all of the requisite parts assembles a working Boeing 747, which, Hoyle and others believe, mirrors the likelihood of life evolving from total, chance, chemical reactions. I would argue that some parts would need to be screwed on and riveted (I think? I'm many things but an airplane engineer is not one of them), thus making it impossible for the airplane to be assembled. That also is essentially the point of the Junkyard Tornado, that something is so fractionally possible that it should be considered simply impossible. I would say that the proliferation of life as we have it, from newts to psychedelic mushrooms to sumo wrestlers, is vastly less possible than the construction by wind of the Boeing 747; rivets and all.

 Richard Dawkins in *The God Delusion* says that God is the "Ultimate 747" in that He would need a creator to create Him if we're staying consistent with what we know of as time. As Dawkins remarks in his book,

> Intelligent Design suffers from exactly the same objection as chance. It is simply not a plausible solution to the riddle of statistical improbability. And the higher the improbability, the more implausible intelligent design becomes. Seen clearly, intelligent design will turn out to be a redoubling of the problem. Once again, this is because the designer

himself (/herself/itself) immediately raises the bigger problem of his own origin. Any entity capable of intelligently designing something as improbable as a Dutchman's Pipe (or a universe) would have to be even more improbable than a Dutchman's Pipe. Far from terminating the vicious regress, God aggravates it with a vengeance.

Time is a measurement in our reality like length is the measurement of the First Dimension. When you get to the next dimensional level, you begin to experience the unshackling of the rules of the last. Our creator can exist outside of, and before, the Big Bang. In fact, unless everything we understand is nonsense, the beginner of the universe- be it the energy catalyst or God- *must* have existed before and outside of time.

The universe will continue to expand, it is a great work and man is God's masterpiece. A sentient creature that can adore him. But man will never become God. I theorize that man will never create a machine more complicated than the God that created man. The Dutchman will never create a pipe (to use Mr. Dawkins example) more complicated than the Dutchman himself. Nothing created will ever be more complex than its creator. God is the only figure unaffected by the concept of regress. It is the only theory that works, and the only chicken that will ever hatch to cross the road.

Chapter 7

Guilty by Association

Great Thinkers existed in ages before the Internet.

Of course, many of the ideas in this book are not original in any real sense of the word. Men and women, vastly wiser and more intelligent than myself, have thought these things and written them down for the past twenty or so centuries, and in the Christian tradition, we revere great thinkers of the past and recognize their achievements and contributions, regardless of the age in which they were made. There is an argument that gets kicked around a lot in the Atheist circles, and it typically goes like this, "X religious person might have had a point or two here and there, but remember he believed also in Y." There is a logical fallacy at play here referred to as the *Ad Hominem* argument. The idea is that a person's argument is wrong *because* of their past, personal characteristics, or other beliefs *as opposed to* the merits of (or lack thereof) *the argument* in question. Italics can only accomplish so much, take this old joke as an example:

A coach is driving along a city road with his high school football team in the middle of Georgia. In the sweltering August heat the old bus jumps along the road with a full complement of restless teenagers, when suddenly, it blows a tire in the quickly dispersing city sprawl. The truck finally comes to a stop in front of a mental health institute with a fence separating the patients and their nice green lawn with chairs and different activities set up outside. The boys file off the bus while the coach and the driver begin getting the wheel off. The coach set the lug nuts inside of the old wheel and propped it against the sidewalk's curb and one of the older patients walked up to the fence to look at the commotion and tried making conversation with the players.

"Seems like you fellers have quite the problem on your hands." He said with a cackle.

"Ignore him." The coach said quietly to the boys while making a polite nod in his direction. A few minutes later the players knock one of their fellows into the tire, it flips and drops all the lug nuts down a storm drain. The coach swears loudly and throws his hat, now having no way to get the bus home without being able to secure the spare tire in place.

Looking on the old man said, "Hey coach, just take one lug nut off the other three wheels and fasten the spare on long enough to drive to an auto store and buy five more."

The coach and all the player's jaws hung open realizing the brilliantly simple plan.

The old man looked at their stares, confused, and then exclaimed in answer, "Oh they put me in here because I'm crazy, not stupid!"

The coach and players were surprised that the solution came from a crazy man, but the mental health of the patient made his idea no less valid as a solution to the problem.

The *Ad Hominem* argument used against many theological scholars, such as Aquinas and Augustus, is that they could not present strong philosophical ideas based on their lack of modern knowledge in so many "key" areas. In *god is Not Great*, Christopher Hitchens began his argument against the metaphysical aspects of religion by pointing out that Thomas Aquinas (born in 1225) half-believed in astrology, and St. Augustine (born in 354) believed that the sun revolved around the earth. I'm not sure what Hitchens was getting after, but I guess I'll bite. The argument goes like this: a doctor takes his car to a mechanic because there is something wrong with the engine. Turns out it was just a bad spark plug and everyone has a laugh at the doctor's expense. Suddenly a mechanic drops to the floor clutching his chest, and keels over. The Doritos, it would seem after all those years, were having the last laugh. The doctor rushes to help when suddenly the head mechanic stops him from saving the man with a heart attack saying, "You didn't know to just change a spark plug? And you think you could *save that guys life?* Please. He's better off with us."

The obvious problem with the argument is that lacking knowledge in one area obviously has nothing to do with lacking knowledge in other areas. That great philosophical thinkers existed in ages without the internet is beyond question. That we should throw out their ideas of the meta-physical because they don't know about the Andromeda Galaxy is as laughable as faulting Plato for being unable to recite the *Star-Spangled Banner* in English from memory.

I have no doubt that, although we may reject the idea in his books, Mr. Hitchens has a great command of the English language—or a mean recipe for guacamole. That Darwin contributed greatly to the study of animals and plants is beyond question, no matter what your beliefs about the earth or antelopes. He made observations: some have found themselves mortared into the foundation of the natural studies, others we may toss to the wayside. His thoughts on finches were sound, but they were far from complete.

How the beak grows and evolves is something the creationist and the Atheist can find in their common ground. But a problem arises when the clock is turned backward several ages to when the world didn't have finches but was encountering the first fellows to try and fly. Here we find a dilemma that no amount of random mutating then natural selecting can properly solve. We find a gap in the record that no theory, equation, or discovery can truly fill.

Chapter 8

Killed by Clothes

Some biological systems are too complex to have evolved piece by piece.

There are some things stretching back through time that I can remember in such vivid detail that they still make me angry if I spend any amount of time remembering them. Army Basic Training and Infantry school is one of those. In my day it was a 13-week grind where you were transported to Fort Benning, GA, and then cooped up with a group of, again at that time, all young men who ranged between the criminally insane and a guy with a master's in business from Harvard. Fort Benning in August is also one of the hottest places on earth by heat index, given its nearly constant 90%

humidity and temperatures often in excess of 100 degrees. The infantry school is in a place affectionately referred to as "Sand Hill", and this is because the entire area is a dismal series of steep hills covered in sand. As a punishment for nearly anything, and I do mean anything, you and two hundred and fifty of your comrades have to charge up the nearest hill in boots, uniform, holding weapon, and back down. Being last will give you scorn, being first will give you the opportunity to scorn yourself for having volunteered to come to such a place.

A few weeks into the training you are instructed to assemble your bullet proof vest, or the Improved. Outer. Tactical. Vest. The IOTV weighs thirty pounds, does not like you personally or professionally, and is powered by the fleeting remains of your hopes and dreams. On a particularly hot morning, I was introduced to the IOTV by wearing it to an obstacle course. I'm not sure what it was about that day but I was still recovering from getting pneumonia in both lungs a few weeks earlier and I felt nauseous in the withering heat. Add to that, the stifling weight of an

IOTV that didn't come close to fitting my torso, and I was all set to throw a pity party for myself and invite the whole platoon. I was being killed by my clothes. Stabbed in the back by what was on my back. Luckily for me, the resident head case had a psychotic break which drew the Drill Sergeant's attention away from my pouting and I was able to sneak through the worst series of obstacles. I went on to trip so badly over a large wall that my rifle nailed me in the nose and I was almost hung by the sling (it left a mark around my throat, also fodder for the abuse cannon). We all had a good laugh and one drill sergeant fell to the ground, shaking and weeping.

It's a lively debate in the military whether the IOTV makes you more fit to survive or less and I would always argue that the vest makes you less likely because no matter how fit you are, you just can't maneuver in it without feeling like you are wearing a fat suit. And this leads me to my point: the pre-cursor to an evolution must make you more likely to survive, not less.

This is known biologically, as *fitness*.

Some things in nature have an obvious evolutionary delineation. In 1831 Charles Darwin, at the age of 22, accepted a position on board the HMS Beagle as it went on an expedition to the Galápagos Islands. He was originally interested in furthering his Geology and Fossil studies and left many of his duties as a naturalist to a helper, one of which, shooting little birds and preserving them, would earn Darwin undying fame in the West. In England the young scientist made a connection between the obviously similar little birds, and realized that they were nearly identical except for the size and shape of their beaks. Examining the 12 or so different species Darwin made an educated guess: one ancestor bird had begotten several descendants on the different islands that used their beaks for different methods of collecting food. The pecker upgrades that helped were passed on and led to a dozen or so different birds. One finch had a longer beak which allowed it to extract a certain type of grub from holes in trees. The longer beak mutation, even 1% longer at a time, made that finch more fit than others and so the genes kept passing on. These differences led to a speciation of several different birds. Darwin had tripped across one of the greatest discoveries in the field of Biology.

Our friend, Mr. Dawkins, wrote a book about what he calls, "Mount Improbable"—in response largely to the ideas of *Irreducible Complexity*, or the belief that some biological systems are so complicated that they require an intelligent designer to exist. In it, he argues that no matter how "improbable" an organism or adaptation, what is being viewed is the "pinnacle of a long, slow climb," often over millions of years up the "Mountain" of improbability. And of course, this is a clever and clear way of understanding many miracles of nature. As he also excellently pointed out in a TV interview, the human eye may seem Irreducibly Complex when in reality it is built on millions of years of small improvements. The beginning of the eye was a simple light sensing organ that would certainly give an advantage to an ancient ancestor over one who couldn't sense light at all. This is totally reasonable, of course, but does less than nothing to describe how a system built on supporting structures could evolve without one of those supports. The eye, for example, could be that light sensing cell which becomes many cells, which eventually develops a lens of some kind (could be a single bit of moisture which focuses some light on those light sensing cells). This you can see leads step by step over the eons to an eye.

Remember that I said each earlier iteration of an organism's biological make-up must at least make the creature as survivable as it was before the mutation. If this weren't the case, it would likely be selected against. Imagine that human men evolved the IOTV organically. By the time I went through training, most of the Infantry had adopted the common Plate Carrier, which was wildly adjustable, weighed ten pounds less, provided basically the same level of protection, and looked way cooler (if you have ever served you can guess which one of those characteristics made it more widely used). The Wikipedia page for it says that the Plate Carrier was developed by the Army, but considering how sensible it is I'm tempted to correct the page. The fact that the IOTV would one day mutate into the stylish Plate Carrier worn first by all the "cool guys" (Green Berets, Navy SEALs, etc.) until eventually it got into the hands of everybody else. This is possible because an "intelligent" designer in the Army took one aspect of the soldier's kit and evolved it to make the soldier more "fit".

In nature, this same idea is also the case. Except that no Atheist agrees.

Dawkins is fond of lumping the wing in with his list of counter points to typical Intelligent Design scientist's theories and arguments. Dawkins makes the common claim that 1% of the wing is better than no wing for certain critters pre-disposed to scrambling through trees. The ides is that if one were to fall from very high up that that one percent of a wing might so retard the fall from a certain number of feet as to protect the mutated critter, whereas the un-mutated animal would be hurt or killed. This sounds alright—but is nonsense.

Imagine for a second that there was such a creature called a "Squawker". Let's call him ScootsII. ScootsII is a feathery fellow, totally unaware of the nobility of his name. He and his friends eat bugs native to the tops of medium-height trees. ScootsII is born with slightly elongated arms that have slight rear flaps on them. These flaps comprise .1% of a wing and are the first such mutation of their kind.

One day, ScootsII and his best friend Richard were out hunting in the evening, when an eagle flew after, and tried to kill them. ScootsII is .1% slower, let's say, and because of the .1% agility deficit, he trips on a piece of a branch, and falls 22ft onto a rock on the forest floor. Luckily he has those flaps, right?

Wrong. Because ScootsII is an animal. ScootsII has no understanding of aero dynamics or propulsion physics. He was not taught how to flap his flaps in a manner that might so retard his fall, and thus after breaking a few ribs and getting the wind knocked out of him, ScootsII is murdered by the eagle. And thus, the genetic mutation of the wing is lost for that generation of Squawkers.

For an evolution to take place, at least according to the theory, it must make the mutant at least as viable as its parent. This is not up for debate. This is not pseudo-science. This is not even science and I am not a scientist. This is basic logic that is ignored by the Atheist community for reasons we will get into. There is a terrifying amount of anthropomorphizing that is necessary to close the gaps in a world without God. Dawkins likes to point out that, "The forests are replete with gliding or parachuting animals illustrating, in practice, every step of the way up that particular slope of Mount Improbable." I went out to gather firewood once and I saw a medium sized dead tree still moored to the earth. When I got to it, I had to break it free from the last of its moorings when out popped a squirrel with big black eyes and white stripes. I had never seen anything like it and so naturally I stopped what I was doing to have fun with a little forest critter. If I leaned left around the stump it would jump right, if I leaned right it would go left, etc. Finally I took the next logical step and tried to slowly grab it when all of the sudden the little guy turned around, leapt toward another tree and then glided perfectly to it. It was without a doubt the most surprising thing I have experienced in nature and I was delighted.

The friend in the woods, let's call him Hopper, was some form of "flying" squirrel. A simple thought experiment can get us to the top of his own "Mount Improbable". Let "Mount" here equal his effortless gliding ability which he has mastery over. Lets say Hopper began a million years ago as a squirrel ancestor named Hop. Hop was a little critter that ran around trees, ate bugs and bullied Squawkers. Now Hop and his friends had a propensity for jumping. They all did it the same way except that Hop was a mutant born with the trait of having slightly flappier under arms than his friends. When he and the others would jump this gave him 1% of the ability to glide. It made him .2% slower but that negative aspect is made negligible by an extra 2 inch "flight" on a long jump. This mutation makes Hop *at least as viable* as his friends and slightly more fit. Because of this, the probability is high that he will pass on his genes (plus he has a great personality and a winning smile. The babes love him).

500,000 years later, a descendant of Hop has very flappy underarms, and his name is Hops. Hops has a mutated gene which gives him a little extra horsepower for jumping, in addition to the flappiness of his arms. Some kinds of Squirrel ancestors don't have the underarms and instead are faster. They live lower in the forest and can run on the ground longer distances than Hops. Those squirrels decide they don't want to breed with Hops so they eventually become their own species, whereas Hops and friends begin to *purposefully* jump and glide. This key mutation and species trait spur a continuing "doubling-down" on gliding, and it winds up with Hopper. Now, here's the important part:

If, at any point, Hop, Hops, or Hopper started flapping their arms, they would die like ScootsII—to the amusement of everyone not involved. That's because gliding has nothing to do with propulsion. A human can very well wear a wing suit, but try flapping his arms and he will careen into the earth. Humans can make the very best glider, but it will simply ride the air until landing without propulsion. We can easily get up Mount Improbable if we're talking *gliding*, but the second we mention *flight*, God turns round and meets us at the gates with open arms.

The wing is an example of an evolutionary idea called *Exaptation,* or "an evolutionary adaptation which is co-opted for another use." The idea here is that the arm exists already, but it evolves in such a way as to be co-opted—first for extended jumps, and then for flight. The above-example of the leaping critter is the most likely explanation. The next possibilities are less likely: that the wing evolved first for catching prey and/or it evolved from a bipedal creature that was already leaping into the air.

The first is fatally unlikely since catching unwilling prey would require a stouter, and therefore, heavier forearm, with which to take damage in the act of pinning the target. The second would be possible if there was evidence of such a creature, or evidence as to why such a creature would want to jump like that, or even, evidence regarding when a creature ever jumps like that (or possibly some mathematical calculations as to at what point would such an evolution make a creature more likely to reproduce—and at such a rate as to pass on those genes to begin it on the way to flight). But before it gets to flight, it would need a myriad of deep and costly genetic changes—such as hugely decreased bone density, lighter and more enduring muscle tissue, a reworked cardio-vascular structure, and feathers angled and shaped in such a way as to create an air foil. And because that is all impossible to be left up to "chance", we return again to where we started.

The wing, Mr. Dawkins, is Irreducibly Complex. If you ever find evidence of a parachuting animal, please feel free to contact me.

Chapter 9

Atheism: The World's Most Boring Religion

Atheism is a religion which requires especially bold leaps of faith to justify a universe without a known or definitive beginning.

The American Atheists are one of the foremost sects of the belief system and operate a bland website (with no pictures) under the very lucky domain name: www.Atheists.org (so close to a .com). In it, they define their "why" statement:

> "Atheism is not an affirmative belief that there is no god nor does it answer any other question about what a person believes. It is simply a rejection of

the assertion that there are gods. Atheism is too often defined incorrectly as a belief system. To be clear: Atheism is not a disbelief in gods or a denial of gods; it is a lack of belief in gods."

They make the clever analogy that if Atheism is a religion, then not collecting stamps is a hobby. Witty, but wrong.

Timothy Keller, pastor of the viral New York City Church, Redeemer Presbyterian, makes an argument in his book, *The Reason for God*, against the false dichotomy of "religion/faith" and "Atheism/science":

> Scientific mistrust of the Bible began [in the enlightenment] with the New Testament belief that miracles cannot be reconciled to a modern, rational view of the world. Armed with this pre-supposition, scholars turned to the Bible and said, "The biblical accounts can't be reliable because they

contain descriptions of miracles." The premise behind such a claim is, "Science has proven there is no such thing as miracles."

But, imbedded in such a statement, is a leap of faith. It is one thing to say that science is only equipped to test for natural causes and cannot speak to any others. It is quite another to insist that science "proves" that no other causes could possibly exist. John Macquarie writes, "Science proceeds on the assumption that whatever events occur in the world can be accounted for in terms of other events, just as imminent and "this-worldly". So a miracle is irreconcilable both with our modern understanding of science and history."

Macquarie is quite right to assert that, when studying a phenomenon, a scientist must always assume there is a natural cause. That is because natural causes are the only kind its methodology can address. It is another thing to insist that science has proven that there "can't" be any other kind. There would be no experimental model for testing the statement, "No supernatural cause for any phenomenon is possible." It is therefore a philosophical pre-supposition and not a scientific finding. McQuarry's argument is ultimately circular. He says that, "Science, by its nature, can't discern or test for supernatural causes and therefor those causes can't exist." The philosopher, Alvin Plantinga, responds, "McQuarry perhaps means to suggest that the practice of science requires that the scientist reject the idea of God raising someone from the dead. This argument is like the drunk who insists on looking for his car keys only under the streetlight on the grounds that the light was better there. In fact, it would go the drunk one better: it would insist that *because* the keys are hard to find in the dark, they *must* be under the light."

The other hidden premise in this statement, "miracles cannot happen" is, "There can't be a God who does miracles." If there is a creator, God there is nothing illogical at all about the possibility of miracles. After all, if he created everything out of nothing, it would hardly be a problem for him to rearrange parts of it as and how he wishes. To be sure that miracles cannot occur, you would have to be sure beyond a doubt that God didn't exist. And that is an article of faith.

When I was a senior in college I coached high school sports, volunteered with the popular youth outreach, *Younglife*, took 21 credit hours, had military duties to attend to, and was training six-days-a week for my coming enlistment. I was, in a word, busy.

I majored in History, and was always working on papers of one class or the other, so when it came to my final general education class of Biology 102, I typically sat in the back and read or wrote, while the professor kindly went on about whatever it was he was lecturing about. Up until that point this was my idea of evolution: a fox is born in North America. It migrates more northerly, where it's colder. A really chilly winter comes along, and the fox swears a lot as it endures the cold. Its own cells then remark, and I'm paraphrasing, "Jeepers Creepers its freezing out here!" And through the generations, baby foxes are born with slightly more and denser fur.

In class that fateful day, we got through Evolution. Somewhere along the lines, I heard what they were saying about the "random nature of certain traits" that are inherent to evolution. My hand shot up and we talked about obvious instances of design, which I thought Atheists agreed with, but which they just attributed to cellular strategizing on some level. Nope. "All random mutations are selected for and against by the environment." The ones that are helpful get passed, the ones that are harmful, get eaten. I was incredulous. I had always assumed non-theistic evolution was nonsense, but I expected more substance. I expected better explanations. But there are no "better" explanations. And this gets at the heart of what I'm trying to say:

The deeper you dig into Atheism, the less you find.

In *Darwin's Dangerous Idea,* Daniel Dennett compares the fundamentals of Creation and Atheistic evolution as the difference between a Sky Hook and a Crane on the ground. He out of hand rejects the notion that certain devices that are Irreducibly Complex (the murderous mind control of our dear scoots, the wings that might have saved ScootsII) could simply have been made by an intelligent designer from outside the world. He argues that this is basically a Skyhook. The idea is that it descends on the world from the heavens and interferes with the natural processes in such a way as to gain a positive result by its own determination. Creation, he argues, is like explaining the building of a tower by using a hook that comes from the clouds. It grabs the pieces and picks them up, but you do not *know* what that hook is tethered to. He then makes the case that natural selection and evolution are a crane: we can see their base and their operations, and most importantly in his view, the materials from the crane come from the environment, all must be a part of the world that it exists within. This idea is widely cited and the terms, "Skyhook" and, "Crane" are regular fixtures in the jargon of this subject.

Three problems spring up immediately with the logic of this argument. Firstly, what or who built the crane? Go ahead and have your brain open its pocket or purse and remember what I said about the Second Law of Thermodynamics: all matter tends toward disorder. Systems deteriorate, fires go out, energy dissipates. A crane can't make systems more complex than the crane itself. Computers can be made by men to have more computing power than men. But you will never find a computer that can create songs, while healing itself from a wound attained on a rock climb (by digesting organic material), all while feeling anxious that the other computer it has a crush on will not respond to the text it sent. These are not the things computers do. It's what *we* do.

Secondly, I reject out of hand the notion that the "builder" of the world must exist "within" it, like a crane. A painter is never the same as his painting. Even if one of those body-painting weirdos walks down the street naked, except for bearing some "artistic" depictions on his skin, he's not the same thing as a painting. He still created upon a "canvas", which in that case and unfortunately for parents everywhere, is his naked body. An architect can go inside his creation, but he is not the building. A great sculpture may come from the imagination of a sculptor, but the rock cut into shape is something entirely different in every way from the man who worked the stone. The Skyhook, as we see, is the only device that works.

Lastly, we can get into what I think is the most pressing question: why in blazes aren't we asking about how the *ground* under the crane got there? Certainly this metaphor falls apart when we ask for cranes and hooks and levers and pulleys to describe the building of the stars and planets and gravity; but that, of course, is the real point of this chapter. How did anything get where it started to become what we now know?

And the answer is…no one knows. Many feel that this is a more satisfying answer than the Creationist and Intelligent Design explanation of "We do not understand the workings of God, and likely can't understand them in their totality". But the more you dig into Atheism, the less you find. When pushed on the issue of *how* did the universe start, Richard Dawkin's best answer is this: "We don't yet have an equivalent crane for physics. Some kind of multiverse theory could, in principle, do for physics the same explanatory work as Darwinism does for biology. We should not give up hope of a better crane arising in physics, something as powerful as Darwinism is for biology." *The Multiverse Theory* is a lot like Middle

Earth and The Force: its totally fictional. The Theory is, in its most basic form, simply states that our universe came from another one. It is a staple in the Science Fiction Genre. Its only ever cited in non-fiction when all other explanations have been exhausted, or, in the case of Intelligent Design, dismissed out of hand.

Dawkins cited the theory in this example to say that the "material" for the beginning of the Big Bang would come from a different dimension or universe. As we saw before with Sean Carroll, that satisfies the fatal question of "how" *we* got *here* but it does nothing for the now more fatal question of how *that* universe got *there*.

The further down the staircase you go inside the house that Atheism built, the weaker the materials become. Eventually, you realize that there is no foundation. The whole thing is a sham. And the rejection of absolute truth in the universe is killing us. The Atheist ship took off without oars. There is no satisfying explanation for non-theistic evolution because one does not exist.

I'd like to make two, rather notable generalizations, consistent with my research and anecdotal evidence: Christians tend to be more proselytizing and convert more in communities, while Atheists tend to have a disproportionate showing among the academic ranks, often converting in isolation, while reading or watching videos. In every regard, Atheism is a religion. It just happens to be boring. So boring, in fact, that it can be sneaky, and can get away with statements as ridiculous as, "Atheism is simply a lack of belief in any god." (Showing that they themselves can't get around using the word "belief") If you think I'm splitting hairs, I promise I'm not. It's vital that we agree on what Atheism is. A few paragraphs down on the "What We Believe" tab of www.Atheists.org it says in bold font, "Atheism is about what you believe. Agnosticism is about what you know."

It is about what you believe. In many ways, Atheism's creed is much like the proliferation of Norse mythology throughout the Viking domains. It had no written rules or beliefs. There were in those days only accepted cultural ideas and a few basic stories that knit their religion together loosely. They agreed on several deities, how everything began, and how everything would end. But those are all rough approximations and would have depended mostly on the speaker at the time.

They really just had some certain ideas and opposed the beliefs of other religions—mainly in that they liked the concept of Valhalla. The certainty of their after-life gave them a feeling of control over their fate: live a worthy life, end with a warrior's death, and spend eternity fighting and partying with the gods until Ragnarök. In a similar way, I think Atheists like the idea of, "The great general anesthetic," as Richard Dawkins refers to death, because it gives a certainty about an issue which is terrifying. To a lesser degree, soldiers draw great comfort in knowing when they will return home from a deployment, when a war will end, or when the training exercise will be over. Privates are famous for their propensity to speculate on changes in the weather or geo-politics and often resort to interpreting body language of officers and authority figures in which to discern the inevitability of the end. In the case of a deployment or a war, the end is a comforting thing and gives the soldier the ability to compartmentalize everything that will happen before then. In the case of an exercise, he can only be so cold and hungry for so much longer before he can return home. The necessity of an end gives control where there is none to be had.

However, it is an illusion. As any veteran would tell you, it can always be worse. The war can always evolve, the deployment can always be extended, and the Lieutenant was chosen for duty because of a unique propensity to get lost in the simplest of terrain. To believe definitively one thing or the other is a way of attempting to exert control over the chaos of being. To rule God out is to also rule out a cosmic judgement for your or others' actions. To rule Him out is to say there is no "x-factor" in the lives of men to further complicate the world.

Chapter 10

Cover and Move

To reply, "Because there is no God" is to create more questions than you answer.

"I'm up!" I yell, as I awkwardly push myself up and charge across the sand on malnourished and tired legs. It is the fourth week of my basic training in one of the hottest places on the planet earth that week. My mind is awash with swearing and self-accusation.

"He sees me!" I scream with a little cough. My lungs still hurt from pneumonia. *Why am I here?* I wonder to myself, mind wandering to a meaningful career of youth ministry that could have been. *Women. Nice, civilized people. Intelligent conversation. Mental stimulation.*

"I'm down!" I bellow as I collapse into the soft sand, rifle bobbing underneath me. *Air conditioning. Shorts and a T-shirt. A Chipotle burrito for dinner. Coco Puffs for dessert.* My face pushed sand out of the way while my arms pulled along the ground, inching slowly toward the Drill Sergeants. My comrades were "firing" at an imaginary enemy. Their fire was "covering" my movement toward our objective. What we were doing was practicing the basics of infantry combat, often affectionately referred to as "cover and move". In this case your friend, X, fires his rifle at the enemy, Y, while you move to the unprotected side of the enemy who is ostensibly hiding behind cover, like a wall. The bullets present an immediate threat that causes Y to duck behind cover and deal with the imminent threat X produces. You then open up the second you get a clear shot from the new vantage point; this either neutralizes Y, or forces him to deal with you, either by returning fire or by retreating to more cover. During that process—you guessed it—X has already moved to a new attack position, and either ends the enemy or causes him to react. This process is called maneuver, it is the dance of combat, and it is represented at all levels of conflict, from three soldiers shooting on the battlefield, to fleets at sea engaging each other, to armies moving across continents engaging other armies, and so on. It also is represented in the Atheist argument, except in this case, its cheating.

During my research, I came to call it the Great Atheist Shell Game because the whole process reminded me of the original County Fair game, wherein the carney has three half-clam shells. They put a pearl on the table, cover it up, and then move the pearl between shells and you try to follow the movements and correctly guess which shell the pearl is under. The game is also notoriously easy to cheat because as the hands speed up the carney can, by a sleight of hand, remove the pearl under the table or into their palm while your brain momentarily gets left behind in the movement of the shells.

Two passages highlight my point: the first is from *The God Delusion*, and is an unedited, intra-chapter section. It is everything between the beginning of his third chapter and the next section. Nothing is edited out, lest I be accused of not giving the Dawkins his due.

The five 'proofs' asserted by Thomas Aquinas in the thirteenth century don't prove anything, and are easily—though I hesitate to say so, given his eminence—exposed as vacuous. The first three are just different ways of saying the same thing, and they can be considered together. All involve an infinite regress—the answer to a question raises a prior question, and so on ad infinitum.

1. The Unmoved Mover. Nothing moves without a prior mover. This leads us to a regress, from which the only escape is God. Something had to make the first move, and that something we call God.

2. The Uncaused Cause. Nothing is caused by itself. Every effect has a prior cause, and again we are pushed back into regress. This has to be terminated by a first cause, which we call God.

3. The Cosmological Argument. There must have been a time when no physical things existed. But, since physical things exist now, there must have been something non-physical to bring them into existence, and that something we call God.

All three of these arguments rely upon the idea of a regress and invoke God to terminate it. They make the entirely unwarranted assumption that God, himself, is immune to the regress. Even if we allow the dubious luxury of arbitrarily conjuring up a terminator to an infinite regress and giving it a name, simply because we need one, there is absolutely no reason to endow that terminator with any of the properties normally ascribed to God: omnipotence, omniscience, goodness, creativity of design, to say nothing of such human attributes as listening to prayers, forgiving sins and reading innermost thoughts. Incidentally, it has not escaped the notice of logicians that omniscience and omnipotence are mutually incompatible. If God is omniscient, he must already know how he is going to intervene to change the course of history using his omnipotence. But that means he can't change his mind about his intervention, which means he is not omnipotent. Karen Owens has captured this witty little paradox in equally engaging verse: "Can omniscient God, who Knows the future, find The omnipotence to Change His future mind?" To return to the infinite regress and the futility of invoking God to terminate it, it is more parsimonious to conjure up, say, a 'big bang singularity,' or some other physical concept as yet unknown. Calling it God is at best unhelpful, and at worst, perniciously misleading. Edward Lear's Nonsense Recipe for Crumboblious Cutlets invites us to 'Procure some strips of beef, and having cut them into the smallest possible pieces, proceed to cut them still smaller, eight or perhaps nine times.' Some regresses do reach a natural terminator. Scientists used to wonder what would happen if you could dissect, say, gold into the smallest possible pieces. Why shouldn't you cut one of those pieces in half and produce an even smaller smidgen of gold? The regress in this case is decisively terminated by the atom. The smallest possible piece of gold is a nucleus consisting of exactly seventy-nine protons and a slightly larger number of neutrons, attended by a swarm of seventy-nine electrons. If you 'cut' gold any further than the level of the single atom, whatever else you get it is not gold. The atom provides a natural terminator to the Crumboblious Cutlets type of regress. It is by no means clear that God provides a natural terminator to the regresses of Aquinas. That's putting it mildly, as we shall see later. Let's move on down Aquinas' list.

4. The Argument from Degree. We notice that things in the world differ. There are degrees of, say, goodness or perfection. But we judge these degrees only by comparing them with a maximum. Humans can be both good and bad, so the maximum goodness cannot rest in us. Therefore, there must be some other maximum to

set the standard for perfection, and we call that maximum, God. That's an argument? You might as well say, people vary in smelliness but we can make the comparison only by reference to a perfect maximum of conceivable smelliness. Therefore, there must exist a pre-eminently peerless stinker, and we call him God. Or substitute any dimension of comparison you like, and derive an equivalently fatuous conclusion.

5. The Teleological Argument, or Argument from Design. "Things in the world, especially living things, look as though they have been designed. Nothing that we know looks designed unless it is designed. Therefore there must have been a designer, and we call him God". Aquinas himself used the analogy of an arrow moving towards a target, but a modern heat-seeking anti-aircraft missile would have suited his

purpose better. The argument from design is the only one still in regular use today, and it still sounds to many like the ultimate knockdown argument. The young Darwin was impressed by it when, as a Cambridge undergraduate, he read it in William Paley's Natural Theology. Unfortunately for Paley, the mature Darwin blew it out of the water. There has probably never been a more devastating rout of popular belief by clever reasoning than Charles Darwin's destruction of the argument from design. It was so unexpected. Thanks to Darwin, it is no longer true to say that nothing that we know looks designed unless it is designed. Evolution by natural selection produces an excellent simulacrum of design, mounting prodigious heights of complexity and elegance. And among these eminences of pseudo-design are nervous systems which—among their more modest accomplishments—manifest goal-seeking behavior that, even in a tiny insect, resembles a sophisticated heat-seeking missile more than a simple arrow on target. I shall return to the argument from design in Chapter 4.

First of all, an omnipotent and omniscient God can only be right. He would never need to change His mind, except in charade to teach us apes a lesson. He would know all that He will do as if reading it in a book. It's a cute device, but one so obviously fallacious that it gets to my broader point. Notice for a moment and go back if you must, that he does not answer the first three "proofs". He only acknowledges they are a "dilemma", and then claims it is the problem for God as well. But obviously it's not, as we saw in Chapter 6. That is because he knows (as he admits) the prevalence and ever-green nature of St. Aquinas and his Five Proofs. Dawkins takes the pearl of regression and slips it away. When you guess the shell, like magic, it's suddenly not there. He does not actually attempt to answer your argument at all.

To the fourth proof, Dawkins compares the meta-physical nature of good and evil with physical sense. These are totally dissimilar, and one relies on our perception and opinion (good and evil), the other on sensory nodes. And as far as something that is so smelly it can cause the nose to quit functioning, we can mostly attest. I once forgot about a bag of gym clothes in the locker at school and returned a week later. I didn't realize what they were at first and the smell had a similar effect on my nose as the sun has on your eyes.

The fifth argument Dawkins shares is the subject of the first half of my book. That he would strongly disagree, I have no doubt, especially since much of the beginning of this work is a primer on the most poignant counter-arguments to Atheism and that many of these ideas have existed in some form in other works. But notice how he never actually addresses the fact that certain evolutionary processes (the wing, the parasites that killed Scoots, etc.) obviously show the trajectory of an aimed technology reacting to outside influences over the ages. At the very end of the section, he remarks that we will revisit the ideas in chapter 4. Except he never does in chapter 4 or anywhere else. He also stated the Five Proofs—and then ran nearly as far from them as possible. He put the pearl under the shell labeled "regression", then switched the pearl to the shell labelled "design", and before both could be examined the pearl was slipped off the table and into his hand.

In *god is Not Great*, Christopher Hitchens begins his siege on the study of faith within religion with his fifth chapter, "The Metaphysical Claims of Religion." He writes:

> I wrote earlier that we would never again have to confront the impressive faith of an Aquinas or Maimonides (as contrasted with the blind faith of millennial or absolutist sects, of which we have an apparently unlimited and infinitely renewable supply). This is for a simple reason.

Faith of that sort- the sort that can stand up at least for a while in a confrontation with reason- is now plainly impossible. The early fathers in the faith (they made sure there would be no mothers) were living in a time of abysmal ignorance and fear.

I wonder, then, if anyone had told Hitchens how, in the age of information and secularization, Christianity's growth has become so explosive? According to Georgetown University and a study by their Berkley Institute, the *illegal faith* in *totally secular* Communist China will reach 230 million members—just counting Protestants—by 2030 (which means that there will be more Protestants in China than there are Christians in all of North America). I wonder if Mr. Hitchens considered that these people, educated in a completely secular environment as mandated by the State, are a good data point?

According to a study by Harvard and the University of Indiana and published in the journal

Sociological Studies, only "moderate ("luke-warm" in the Christian parlance) faith is on the decline in America." We could argue that these people were never more than agnostics in practice, and we're likely just seeing a better reflection of what people actually believe, in the polling mega-data. The study, however, contradicts a favorite talking point in New Atheism, and—that is the trend of secularization in the United States. According to the abstract:

> Recent research argues that the United States is secularizing, that this religious change is consistent with the secularization thesis, and that American religion is not exceptional. But we show that rather than religion fading into irrelevance as the secularization thesis would suggest, intense religion—strong affiliation, very frequent practice, literalism, and evangelicalism—is persistent and, in fact, that only moderate religion is on the decline in the United States. We also show that in

comparable countries, intense religion is on the decline or already at very low levels. Therefore, the intensity of American religion is actually becoming more exceptional overtime. We conclude that intense religion in the United States is persistent and exceptional in ways that do not fit the secularization thesis.

The pearl here is taken and put under the shell marked, "Great Thinkers," then, quickly moved to "immaturity and bygone beliefs," until finally, after the moving and shifting of shells, the pearl is taken off the table. There was never going to be an actual refutation of Aquinas's real beliefs. Only silence in the face of deep mysteries, still left unanswered by a belief system that does not include God. The best an antagonistic argument can do against the Truth to shift away any responsibility to prove a fact, one way or the other. It is, unfortunately, not quite so simple as that.

Chapter 11

The Burden of Proof

Religion exists everywhere because people naturally react to the stimulus of the invisible God.

Bertrand Russell, the renowned British philosopher and one of the key founding fathers of modern Atheism, is famous for, among other works, a paper that was commissioned in 1952 by *Illustrated* Magazine. It's one of his most cited ideas, and the thought experiment is called "Russell's Teapot":

> Many orthodox people speak as though it were the business of sceptics to disprove received dogmas rather than of dogmatists to prove them. This is, of course, a mistake. If I were to suggest

that between the Earth and Mars there is a china teapot revolving about the sun in an elliptical orbit, nobody would be able to disprove my assertion, provided I were careful to add that the teapot is too small to be revealed even by our most powerful telescopes. But if I were to go on to say that, since my assertion cannot be disproved, it is intolerable presumption on the part of human reason to doubt it, I should rightly be thought to be talking nonsense. If, however, the existence of such a teapot were affirmed in ancient books, taught as the sacred truth every Sunday, and instilled into the minds of children at school, hesitation to believe in its existence would become a mark of eccentricity and entitle the doubter to the attentions of the psychiatrist in an enlightened age or of the Inquisitor in an earlier time.

In 1958 Russell elaborated on his rebuke of the Theist argument:

> I ought to call myself an agnostic; but, for all practical purposes, I am an Atheist. I do not think the existence of the Christian God any more probable than the existence of the Gods of Olympus or Valhalla. To take another illustration: nobody can prove that there is not, between the Earth and Mars, a china teapot revolving in an elliptical orbit, but nobody thinks this sufficiently likely to be taken into account in practice. I think the Christian God just as unlikely.

Russell's is a convincing argument, or it would be if he made *one* argument, but instead here he makes two and combines them in the same paragraph before passing them off as the same, when in reality they are apples and oranges. In the first, he is putting himself in the place of introducing a new idea that has never before been conceived, and presenting it to the public for them to judge it. We all agree it to be a silly idea. He then takes his invented analogy, makes it a strawman, and introduces it to the next argument—which is tailored to be like a generic religion.

There are several problems here. One, We could make Russell's Teapot work as an analogy, but doing so would rob it of any argumentative usefulness. To accomplish this, I would make the following changes to his example: say that we had some form of sensor between the Earth and Mars that tracked small objects. We put it there for one scientific reason or the other, and one day it picks up a strange anomaly in the form of a teapot. So many months later, it registers the same anomaly. Down to the day, the same number of months later, that teapot is registered again. It becomes obvious that the object in question is in orbit around the sun. Scientists are then puzzled by this anomaly and wonder what it could be since it does not fit the bill for other forms of space debris. There is now, entered into their community, a question that needs to be answered. The first part of Russell's analogy could then be entered into this intellectual environment, because now someone could present a solution to a question that satisfies the *what* (what could that teapot-shaped anomaly be? What could that anomaly be if it has the same general weight of a teapot?).

The second part of his analogy will not work, due to how it was presented, no matter how we try. The reason is that, for it to work, the majority of people in some tribe or country would have to have accepted that a teapot found its way into space, into orbit, around the sun. The teapot would have had to accomplish this long before his argument takes place, in order for there to be an organized belief system around it and for there to be books about the belief. The problem is what I described earlier: people can be gullible enough to believe many ideas, but certainly not *anything*.

Two, that Islam, for example, when introduced was an amalgamation of several beliefs already in motion in the Middle East at its conception. Muhammed, assuming he was the one who began the faith, espoused ideas that answered questions that all people of the region already had. It filled in the blanks the of different *why* questions (why are we here, for what purpose were we created, why is their evil in the world, etc.) and answered them to relative satisfaction of the local populous. In a sense, Islam had (and for many, has) a certain undeniable utility that was naturally selected for in the ideological battlefield, whereas other beliefs and religions at the time were selected against. This is the opposite of Russell's assertion which is that the religious invent problems which only they can "fix".

Years ago, when I was a junior in college, my friends and I from Younglife used to meet at one of the girl's houses to plan out our following week. One day, as we all sat on the couches and were going through some logistics, my friend Phil reached behind himself and mindlessly scratched at something on the couch. It made a weird noise, but because my other two friends (Joe and McKenzie) were facing Phil, we instantly connected the noise to his fingers and the couch.

In the meantime, my friend Abby, lying on the carpet and reading from her computer, didn't make the connection. Somehow, the sound seemed ominous to her out of context and she looked up quizzically at first, asking, "What was that?"

Phil, bored by being serious for more than ten minutes, immediately decided on shenanigans and with a straight face and a twinkle in his eye replied, "What?"

The three of us immediately suppressed smiles and looked at Abby with blank stares and she went back to reading. A minute later Phil made the noise again, but much louder. Abby looked up quickly, obviously distressed, saying, "That noise! Did no one else hear it?"

"Abby. Are you ok?" McKenzie, that beloved sister of mine and partner in crime, said with such an honest concern it frightened me as I went through all the good times I remembered with her and realized she could lie this easily.

Phil made the noise again as soon as Abby went back to what she was doing and she shot up onto her feet, "How can you not hear it?" She demanded, now distraught.

Joe, never one to be left out of an opportunity to make people laugh, was obviously deep in thought with how he could capitalize on this. Phil mimicked concern.

I made an obvious look to the other three as if we were all worried, "Abby have you been hearing stuff lately? Getting enough sleep?"

Phil made the noise again and she jumped, looking right at him but now convinced there was something else wrong. She looked up at the ceiling with doubt, "Do you think it's a… creature?" We all burst out laughing and I rolled onto the floor holding my sides. Phil nearly cried as he turned over and scratched the couch in front of her and she said, "Phil! Did everyone know?" To which we all just nodded in between gasps. It was not the last time we would laugh at her expense, but it is certainly one of the most memorable.

In my story Abby was presented with a question: "What is that noise?". She became disturbed because no one could corroborate what she was experiencing. As she continued to hear the noise and confirm that she was really hearing it, even if others were not, she began to start trying out pegs to fit into the proverbial "square hole". As she tried out different ideas she came finally to the one she asked, "Could it be a creature?" She was really hearing the scratching of fingers on fabric, but she didn't arrive at that conclusion, so she came up with a different one. If this story were like the point Russell tried to make, it would play out like so: Abby is laying on the carpet, stands up, looks at her ceiling and declares, "There is a creature in the attic." Far from being funny, it would be confusing, because we have no reason to think there is a creature in the attic and would then go about testing her claim. Russell's Teapot is an argument that takes unlike ideas and sets them together for the sake of creating a strawman. But as long as we degrade ourselves with cognitive distortions and make untenable arguments, the intellectual coliseum will be dominated by the loudest voices, not the most reasonable ones.

The Burden of Proof is the concept that the one who introduces an idea must also be the one on whom the burden to prove the idea's truth, rests. If I were to tell you that there are actually fairies all around you and you can't see, hear, or sense them in any way, you would tell me I was crazy. If I then told you that it was your job to prove me wrong, you would think I am crazy and foolish. The main argument in all of my research is that the Atheist would say to the theist, "I do not believe there is no God. I simply lack a belief in Him. And anyways, it's not on me to disprove your claim, but on you to convince me." That would be a logical proposition. But let me add to it my own: somewhere along the lines of the great expanse of time, man evolved to such a state that he encountered the *Uncanny*, as the Christian apologist C.S. Lewis describes in his classic *The Problem of Pain*:

> The first of these [proofs] is what Professor Otto calls the experience of the Numinous. Those who have not met this term may be introduced to it by the following device. Suppose you were told

there was a tiger in the next room: you would know that you were in danger and would probably feel fear. But if you were told 'There is a ghost in the next room', and believed it, you would feel, indeed, what is often called fear, but of a different kind. It would not be based on the knowledge of danger, for no one is primarily afraid of what a ghost may do to him, but of the mere fact that it is a ghost. It is 'Uncanny' rather than dangerous, and the special kind of fear it excites may be called Dread. With the Uncanny one has reached the fringes of the Numinous. Now suppose that you were told simply 'There is a mighty spirit in the room', and believed it. Your feelings would then be even less like the mere fear of danger: but the disturbance would be profound. You would feel wonder and a certain shrinking—a sense of inadequacy to cope with such a visitant and of prostration before it—an emotion which might be expressed in Shakespeare's words 'Under it my genius is rebuked'. This feeling may be described as awe, and the object which excites it as the Numinous… …Most attempts to explain the Numinous presuppose the thing to be explained—as when anthropologists derive it from fear of the dead, without explaining why dead men (assuredly the least dangerous kind of men) should have attracted this peculiar feeling. Against all such attempts we must insist that dread and awe are in a different dimension from fear. They are in the nature of an interpretation man gives to the universe, or an impression he gets from it; and just as no enumeration of the physical qualities of a beautiful object could ever include its beauty, or give the faintest hint of what we mean by beauty to a creature without aesthetic experience, so no factual description of any human environment could include the uncanny and the Numinous or even hint at them. There seem, in fact, to be only two views we can hold about awe. Either it is a mere twist in the human mind, corresponding to nothing objective and serving no biological function, yet showing no tendency to disappear from that mind at its fullest development in poet, philosopher, or saint: or else it is a direct experience of the really supernatural, to which the name Revelation might properly be given.

That feeling I would bet, based on the statistics, you are at least familiar with. Ancient men sought to answer many questions, but the primary one was that of purpose. It was to name *uncanny*. It was to order the cosmos. It was to put handles, as it were, on that objective truth at the heart of the universe. The fact that 100% of students are taught something that only 4% end up believing in any meaningful way is not proof of the ignorance of the many, but of the few.

You are, right now, much more comfortable and healthy than anyone in human history before the year 1900. Your quality of life exceeds that of any royal, emperor, or tyrant that lived before the wide proliferation of electricity and modern medicine. You also live in a world with drug overdoses, terrorist massacres, nuclear weapons capable of making the planet uninhabitable, and governments who, at times, aspire to attain a level of tyranny left open for only the most wicked conceivable version of God.

We are vastly more prosperous, and profoundly less satisfied.

Now, consider that 96% of you at least believe God *probably* exists and go another step further with me. As a man living thousands of years ago, you had sex and the likelihood is high that your partner would die, possibly with the child, during one of her many pregnancies. Germ theory was millennia in the making and nearly any major wound was a sentence to die in agony. Starvation was always a few bad decisions away. Tyrants ruled petty bands of murderers and killed with impunity. There is no dentistry, gynecology, toilet paper, or Tylenol. The life expectancy is so low it is laughable. Humans regularly expose deformed babies to the elements to spare them the pain of the inevitable savagery of that ancient world. Now enter this, that a man one day says to his friends as they lay down looking at the heavens, "I bet a God made all those stars." His friends silently agree. Then he says, "And I bet that God is nice." His friends tell him to go jump in a lake. If there were no *uncanny*, that is how this conversation would play out every time it sprung up across the millennia. It is such a bizarrely silly idea it could only take root in human civilization everywhere if we, as a species, were all reacting to a similar stimulus.

    That ancient humans could chalk up the influence of the wind and sun to magic and the workings of the Gods is only to say that they didn't know any better about science and astronomy. But that people all over the world anthropomorphize those gods and continued to do so to such a prolific extent, should deeply unsettle the Atheist reader. In the words of C.S. Lewis, "Remember that all this happened in an age without chloroform."

Before man invented the requisite technology in which to detect sound, we knew that birds made pleasant songs because we could hear and witness the birds singing for ourselves. This phenomenon is witnessed all over the world and when men witness it they ascribe certain anthropomorphisms like, "The birds sing *because they enjoy it*." Or "The birds sing *so that they can communicate across the woods and sky*."

If one day a group of people who can't hear anything come along and say, "Birds don't sing, and we know what singing is because when we're near someone else we can detect the vibrations of their voice." If you wished to correct them you might say that their ears do not work so who do they think they are to tell you what noises birds do and do not make? Or, more maturely, you might bring them to a pet store to listen for the tweeting vibrations of a few finches. If they were afterwards unable to sense the vibrations, you might point out the moving beaks and slightly depressed chest.

At this point, our deaf friends could relent, or they could say, "Well never mind that, those finches are taking deep breaths from all the flying about they do. You have proved nothing. And since I'm the one who doesn't believe in the bird song, you must be the one to describe it to me. The *burden to prove the song rests on you*."

This story, as you might have guessed, is an analogy for the way many Atheists deal with the phenomenon of Theism—something that every major and minor people group has professed a common belief in since time immemorial. This argument is also a common logical fallacy in that I'm appealing to consensus. But logical fallacies are not always fallacious. In the example, I proposed that our deaf friend's physical clues to discern birdsong. My suggestions were rejected as being, for him, *a leap of faith*. He has no way of hearing the birdsong of course, but the fact that he can't hear it does less than nothing to affect the realness of the song. And *song*, by the way, is a traditional means of referring to the interaction of God in the lives of Christians. As if the gentle humming that began in Creation fills all the world.

Ash Dykes, a legendary Welsh adventurer (and a man living out what I wish I could do), described his trek across the breadth of the Gobi Desert. At night, so far away from the constant din of the modern world, totally separated by all the machines of man, he could hear, on a windless night, the hum of his internal mechanisms. It is a phenomenon known to occur only in places so destitute of people and fauna that with no breeze at all you can begin to hear the subtle tick of the body's clock. But if your knee jerk reaction is to disbelieve that account, I want to ask: are you under the impression that your body does not make sound? That the digestion of food, expansion and contraction of the lungs, and steady drum beat of the heart do not produce a low din, so low as to be imperceptible in all but the quietest corners of the world?

The same can be said of a Creator God. That the world was made, I have no doubt and am led to believe by the abundance of scientific facts. That that creator is the Judeo-Christian Yahweh, and further, that His son, Jesus, is who He says He is, I have discovered a profound belief by the weight of many clues and facts, some of which are perfectly anecdotal, based on my own life experiences and interactions. Many are based on facts. The level of evidence, to me, is overwhelming.

That many Atheists claim the burden of proof lies with the believer is sound reasoning—or it would be if it were not for a mighty consensus, continuously reacting as little "radars" all over the world to the *uncanny*. That God is real, we all agree. Who God is, and what He wants, is a totally separate matter and beyond the scope of this book. If the Atheist can't "get God" as it were, by shifting the responsibility to prove facts, she will be left to try and get the whole argument cancelled on a technicality.

Chapter 12

The P Word

To reject God is not a rational decision but an emotional one.

When I was in the 9th grade, I enjoyed a rare respite from all the problems of my childhood. It was a very pleasurable season, smashed in between a rough time in Junior High and an extremely hard year in tenth grade. I went to a small private school in a sleepy and forgotten part of the world, seemingly unbothered by the troubles of society. The grades typically had around ten to fifteen kids and we had all recently left behind the viciousness of the early teen years and were the best of friends. I had several teachers that had known me since I was five and it was as safe an environment as a kid could hope for. When my father went to Afghanistan I was asked regularly about his well-being by one of the coaches, the secretary knew the phone numbers of several extended family members, and the lunch lady could remember the funny way I ate when I was in the first grade. It was in this padded environment that I had developed an intellectual curiosity and a daring to argue with teachers when I "knew" I was right. The overwhelming majority of those arguments I lost because I was, in fact, incorrect, but one sticks out that still rubs me the wrong way.

 I sat in world geography, a favorite subject of mine, and listened while my teacher described the agrarian societies of Dark Age Europe,

 Mrs. B "So you see these windmills? What do you all (She was from Canada and didn't say 'y'all' as most people did in the South) think they were used for?"

 (A few wrong answers later)

Me "They used the wind power to mill grain into flour, among other things (I couldn't think of the other uses but I knew that was right)."

Her "Ok that's a good guess but the answer is more surprising."

Quickly cover the rest of the page. Go ahead and take this time to guess what she thought they were used for. Guaranteed you do not guess it.

Her "They used cattle (I kid you not I remember this as if it happened last week) to turn the big fans on those wind mills to dry out the swampy bogs of their marsh land so they could be farmed."

This woman *honestly thought that windmills were cattle-powered-iron-age-hair-dryers.* She thought that ancient people not only had enough cattle and enough windmills lined up to produce that kind of wind power, but that this was even possible with any man-made machines over hundreds of acres of property.

My jaw hung open while the rest of the class made various noises of consent and wonder.

Me "What? No they didn't."

Her, and she had the nerve to give me an incredulous look, "Yes the ground was constantly too damp to farm, so they had to make more wind with the windmills."

Me "No those huge fans *catch* the wind. The shaft turns huge gears inside the building, the gears are connected to a big millstone which grinds the grain husks and separates out the flour."

And she replied back in a patronizing tone before moving on, while my classmates called me a know-it-all (they weren't wrong, I was a pretty big jackass, but that time I was right) and one friend leaned over and said, "Do you always have to argue with the teachers?"

Me "But you know I'm right. They obviously don't dry the land out."

Him "Ok. So what?"

Ok. So what?

And that is the Atheist question in a nut shell.

Me "Ok but something had to give the potential energy to start the Big Bang. And the material can't have been created, and it can't be destroyed. So there is a problem with a non-theistic creation event."

Them "Ok. So what?"

Pseudo-Science is the great intellectual slur against the study of Intelligent Design. If you can't beat someone's theory, at least make fun of them. If you can't prevail in an argument, you may well have to get the argument cancelled altogether. A common saying from parents to their dorky kids is that bullies make fun of others because they are secretly self-conscious. This is, one, true about bullies, and two, never made me feel better. The prevailing claim among the establishment is that Intelligent Design is Pseudo-Science because it is *unfalsifiable*. A popular example argument is much like Russell's Teapot, and is called Sagan's Dragon:

> "A fire-breathing dragon lives in my garage" Suppose (I'm following a group therapy approach by the psychologist Richard Franklin) I

seriously make such an assertion to you. Surely you'd want to check it out, see for yourself. There have been innumerable stories of dragons over the centuries, but no real evidence. What an opportunity!

"Show me," you say. I lead you to my garage. You look inside and see a ladder, empty paint cans, an old tricycle -- but no dragon.

"Where's the dragon?" you ask.

"Oh, she's right here," I reply, waving vaguely. "I neglected to mention that she's an invisible dragon."

You propose spreading flour on the floor of the garage to capture the dragon's footprints.

"Good idea," I say, "but this dragon floats in the air."

Then you'll use an infrared sensor to detect the invisible fire.

"Good idea, but the invisible fire is also heatless."

You'll spray-paint the dragon and make her visible.

"Good idea, but she's an incorporeal dragon and the paint won't stick." And so on. I counter every physical test you propose with a special explanation of why it will not work.

Now, what's the difference between an invisible, incorporeal, floating dragon who spits heatless fire and no dragon at all? If there's no way to disprove my contention, no conceivable experiment that would count against it, what does it mean to say that my dragon exists? Your inability to invalidate my hypothesis is not at all the same thing as proving it true. Claims that cannot be tested, assertions immune to disproof are veridically worthless, whatever value they may have in inspiring us or in exciting our sense of wonder.

Much of my evidence for *Yahweh*, or the Judeo-Christian God, is a personal faith built upon half a lifetime of observation and interaction, and to me, its an overwhelming weight of clues that led me to affirm the existence of God. But I, of course, would never try to prove that the Creator God was Yahweh with anecdotal evidence (and that is not within the purview of this book, anyway). That Yahweh is the Creator is a *belief* of which I am personally convinced, but it is far from being scientific. Intelligent Design *is* scientific, and of that I seek to prove within the scope of this book.

A scientific claim that I made in the second chapter fits my bill of proof: Toxoplasma Gondii is irreducibly complex. Because of that, it cannot have evolved *unassisted*. As such, I am led to believe that because I have evidence of a created thing, there must, in fact, be a creator *and* an ongoing intelligent designer.

A scientist could take this claim and test it. If it were proven false, then so be it. Back to square one. But if it can't be disproven, it becomes an existential threat to the Atheist world-view. There is a huge problem with claiming that the universe was not created: the theory is a house of cards; if at one place design can be proven, the whole thing crashes down. There can be no hole in the dam that holds back the waters of a Creator. It is why I know, beyond a shadow of a doubt, that a formal refutation will eventually be published on this very topic. That my work will be dragged through the internet gutter, I'm well aware. Because what we're getting at here is the real issue: non-belief in a Creator is not based on facts, but on emotion.

If I were to respond to a good-faith criticism of Christianity wherein a friend said that Christianity seemed impossible and I responded that "God is there you just can't see Him," my friend would be correct to point out that it was conspicuously similar to Sagan's Dragon. But I, of course, am not making that claim. The last two thirds of this book have been a journey to prove that Atheism, far from holding the exclusive rights to science and reason, as is so fallaciously reported everywhere, is a religion that requires leaps of faith and articles of pure belief, often rivaling or exceeding, that of the religious.

The God Delusion, god is Not Great, and *The End of Faith* are cited so heavily in this book because they are seminal works in the study of Atheism and are considered to be some of the most compelling ever written. I have read each several times, and obviously I remain unconvinced, to say the least. This book is meant to "level the playing field," at least to the extent of arguing that there are several large holes in the belief systems that make up the New Atheist movement. That they chide other religions for their "superstitious" beliefs and their "leaps of faith," I find laughable. The argument, "There are many clues for God, but it still requires *faith* to believe and trust in Him." May be initially unsatisfying to our sense of curiosity. But do you know what's even worse? The following argument:

X "I think the entire world came from an infinitely dense, and very small point. It all exploded outward in an instant and very quickly (relative to the quickness or slowness of stars and galaxies) formed into the Universe we know of today and has been expanding outward ever since."

Y "Ok, well how did that ball get there? Was it always compressed or did it compress somehow? On its own? What made it expand? Did something cause that?"

X "Well, we know it did all that because it is the only theory that works in a universe without an intelligent designer."

Y "So you started a scientific theory with an asterisk? That seems intellectually dishonest."

X "Ok. So what?"

Chapter 13

Stalin, Hitler, Mao, Pol Pot

If recent history is our reference, our species is becoming less moral, not more.

While we're talking teachers and studies, another story rings in my ears as I sort through the note cards I wrote out for my research on this book.

I can remember many classroom rants against teachers that involved my heartfelt and revolutionary appeals against the subject being taught for lack of usefulness. Those teachers were easily able to deflect my arguments and fact check my manifestos. No matter my disdain for algebra or chemical equations or poetry, I still had to take the test and, on average, I got a "C".

Then there came my legendary English teacher, a massive New Yorker, who found himself stuck in the stagnant, intellectual swamp of the rural Deep South. At his first teaching job, he was a celebrity for his "Yankee" accent, while simultaneously becoming convinced that not a single student could relate to any of his interests.

He then did the unthinkable for a well-educated man of the North: he studied the WWE. He did so only to win the hearts and minds of his kids, and that leadership lesson has stuck with me ever since. It's probably the only one I can recite from my high school years.

The other thing that stuck with me from that class was the utter brutality of his year-end research paper assignment. A student had to write a *five page* essay. Worse, they had to have *fifty sources*—all on flash cards. I was aghast. I couldn't believe it, and I didn't do it until the night before it was due. It was the pinnacle of all useless skills. An affront to utilitarianism. A mockery of the intellectual endeavor. I might have been right on all those counts, insofar as none of my classmates have likely used those skills in a non-academic setting. I was right about everyone—but myself.

I have used the practice of creating note cards beginning in college and through to the present book you read.

And it's funny, it really is.

In writing my note cards, I would format them roughly thus: a brief description of the passage I'm quoting or paraphrasing, my counter argument, the page number, the book it is from, and the rough theme of the card.

In writing these notes, one theme, in particular, came up across all my research: Atheists love to point out violence in religion; not so much when it involves other Atheists. The summary worked its way down through the first note card from Richard Dawkins' *The God Delusion*, "Dawkins makes special called, 'The Root of All Evils' on BBC about the evils of religion, but does not even stop to consider the one hundred million killed in the 20th century alone by Atheist despots: Stalin, Hitler, Mao, and Pol Pot." To the first book of Hitchens I read, *the Portable Atheist*, where my note card read, "Stalin, Hitler, Mao, Pol Pot." To the end of his first chapter in that book where I had made that same notecard so many times that I just shortened it to, "SHMPP".

When I got to another work of Hitchens (that would be funny if he was not serious), *god is Not Great*, I endured the cognitive minefield of thirteen and a half pages, to find the title: "Religion Kills". I thought to myself, "Ok. Here we go. He will make some obvious statements about Belfast, Al Qaeda, and The Crusades. Surely he will not be so foolish as to leave out all the millions murdered systematically by the Atheist leaders of the 20th century, right?"

Wrong.

The man wrote an entire chapter on the evil deeds done in the name of religion. In what might be my only citation from the Huffington Post, there was a fantastic and short piece written by a Jewish scholar, who rejected the low-hanging fruit that Atheists attack—which is that religion is the cause of all wars, most wars, or many wars. According to this article, Rabbi Alan Laurie states that after compiling all 1,763 known human conflicts in history, "less than 7 percent" were caused by religion—which accounts for less than 2 percent of the body count. Pretty low numbers for a supreme deity (or deities), if you ask me. What about The Crusades? The best estimate is 1-3 million. The inquisition? 3,000.

Compare those numbers to the 35 million killed in just four years because some Austrian got greased by a bunch of anarchists (World War I) and you will start to see what I'm getting at.

Or we could make a much better comparison: the great Atheist political movement of the twentieth century. Communism.

According to conservative estimates, in 100 years, Communists have successfully eliminated over 100 million of their own members. There is something wildly Darwinian about it, that I can't quite put my finger on.

100 million? *100 million!* With those kind of numbers, Communism puts all religions to shame. Those Atheists should make their own league. And they are Atheists, by the way. Marx wrote in his manifesto, "Communism begins from the outset with atheism; but atheism is at first far from being communism; indeed, that atheism is still mostly an abstraction.". And our friend Hitchens would know this, as he included the very chapter in question of Marx, that abomination of a human, in his compilation book *the Portable Atheist.* Stalin, Mao, and Pol Pot were all Atheists. They all massacred millions. If we judge a religion in terms of body counts, Atheism is out in front and it is not even close. But let me here distance myself from Hitchens.

A favorite argument of the Atheist against religion is religion's violent aspects. But religion can't claim the kind of genocidal insanity that Communism can, or Nationalism, Imperialism, racism, greed, selfishness, or even stupidity. And I, being a man who aspires to real intellectualism, do not blame a "lack of religion" for the massacre of innocents. I blame evil and I blame people. Religion is often an expedient tool for the tyrants and criminals, but it is not their teacher. If Mr. Hitchens wants to say that people are evil, than John Calvin of reformed Christian doctrine is awaiting him with open arms. But that is not the point he wants to make. As he restates several times in his second chapter, "Religion Poisons Everything." *Everything? Poisons?* It poisons dead people? Minerals? Orion's Belt? Bacteria in deep sea thermal vents?

That people tend towards being bad, I have no doubt. In fact, we happen to have fancied ourselves the notion that, as a society, we have evolved; that we today are better than our ancestors a thousand years ago. And that might be one of the greatest lies of them all.

In 2008 Oprah Winfrey had a television show called, "Oprah's Big Give." Contestants would vie for victory in a weekly competition-style reality tv show to win 1 million dollars, of which they were supposed to give half away to charity. Every week, money was given to charity, and, for the point of argument, let's pretend that Oprah actually fronted all of that money herself. At the time of the show's release, she was worth approximately 2.5 billion dollars according to Forbes. If she really gave away 1 million dollars of it, that would comprise a donation of 1/2500th of her wealth. According to the 2010 Census, the average American can expect to make over the course of their life, in rough approximation, 1.4 million dollars. Oprah, therefore, has more wealth than the average 1,785 people combined will make in their entire lifetimes. For this giveaway, she made a primetime television show that was advertised everywhere, in order to let the entire world know just how generous she was. And this is not a criticism of her. I have no doubt she does many charitable acts and would take in a stray dog if she found one while driving home. This is a criticism of our society.

C.S. Lewis is fond of using the example of "rest and indigestion" as common reasons for our good actions. We go around patting ourselves on the back for giving a few dollars to help someone who has nothing and attribute it to our goodness. Or we hold our tongue when we could have really let the other guy have it. But is it really "goodness"? Is it that we're so decent, or that we have many dollars to spare, and are well rested and fed. What if we felt the pinch of poverty or awoke with a migraine? What then becomes of our "better nature"?

In the book, *Ordinary Men*, a group of police officers were given instructions to aid in the Nazi "Final Solution". These were not Nazis. They were middle-aged working-class citizens and members of a reserve police battalion. Their commander made it clear to all five hundred that at any time they could return home and opt-out of duty. Only twelve ever decided to do so. The remainder committed atrocities that leave the reader nauseous. They rounded up thousands of Polish Jews, executing many of them one by one. What makes the account so striking is that, as the title implies, *these were ordinary men.* They were subject to the same societal pressures, the same sense of tribe and duty and friendship and family. They had all the normal prejudices and fears. They likely enjoyed drinking beers with friends and if they happened to cross your path after you suffered a flat tire I'm sure they would be inclined to help you change it. And yet, in the course of a war they would drag a pregnant, Jewish woman out in the snow and shoot her in the head. That I pat myself on the back for giving a few dollars to someone obviously in need is only the out of my extreme abundance. It can hardly be called "generosity".

I'm what many people would consider lower middle class, and yet I'm more comfortable than *any* human born before the turn of the 20th century. Oprah Winfrey lives a quality of life *unimaginable* to what the ancients would have ascribed to the conditions of Mount Olympus. The Atheist spends an inordinate amount of time talking about bygone religion and spiritual sentiment. The entire first section of Sam Harris's *The End of Faith* is on the existential threat posed to mankind by the extreme strictures of religion. But men like Harris are also assuming that we're evolving as people. I couldn't think of an idea so unsettlingly naïve as to believe that I'm truly more kind than my ancestors were. More foolish and soft, probably, but more *ethical*, I highly doubt.

Remember that the Black Plague of the 14th century killed anywhere between 75 and 200 million people; in six years, World War II engulfed the lives of approximately 60 million. A war which officially ended by the indiscriminate nuking of two largely civilian cities. And Communism was just getting started at that point. About to take the murder baton from the Nazis, as it were. If the last 120 years are proof of anything, it is that we are regressing as a species, not improving.

Chapter 14

For Which We Have No Answer

Some actions are so selfless that they hint at the nobility of our origin.

On the thirteenth day of January 1982, Arland D. Williams Jr. shuffled down the cramped rows of seats to rest next to the window in the rear of the plane. He kept his coat on, still cold from unusually bitter conditions in Washington D.C. that day. Williams was a banker, and a divorcee with two sons. He was giving it another go and engaged to be married. He was well liked by most, and well-to-do.

The Boeing 737 taxied on a snowy runway, and began it's take off. As the ground fell away, Williams felt a rock in the cabin, and noticed that the ground had stopped falling. The plane leveled out and started to tip forward, then began losing altitude, as panic fired through the cabin. The bleak gray of city traffic in winter closed in on them, as the aircraft struck the 14th St bridge, killing 70 of the 76 people on board instantly. The plane continued off the bridge—and into the icy blackness of the Potomac River.

As the main body broke away under the ice, and as the bone chilling water blasted up towards them, Williams suddenly clicked on. He grabbed his fellow wounded passengers and began ferrying them through the water to float next to the bobbing tail section. He rescued the remaining five members and helped them hold on for their lives. Motorists above called police and the coast guard, but given the ice and conditions, Williams' fellow Americans were unable to help. A Coast Guard rescue helicopter eventually got close enough to start picking out survivors from the water. Clearly the most alert, they tossed Williams the rope first.

He passed it to the woman on his right without hesitation.

She was pulled out of the water and into safety. The helicopter came back, and again, threw a rope out that survivors could cling to, in order to be dragged across the thin ice and to the shore. Williams gave away his spot again and three more men were saved.

Finally, the helicopter came back to rescue the two, remaining men, when Williams again, passed the opportunity to save himself and stayed with the aircraft while the helicopter unloaded its hypothermic passenger on the shore, then began its last attempt to rescue Williams. But in those few minutes the tail section suddenly lost its buoyancy—and dragged Williams under the water.

Let me ask you a question: what did you feel when you read that? I got teary-eyed, goosebumps ran down my arms. Why?

Because a nondescript, middle-aged father—with everything to live for—traded his life for five people he had never met. And for no reason other than pure righteousness.

Sam Harris is a leading thinker on the topic of biological morality as a defense against the idea of righteousness. That people would exhibit the kinds of selfish behavior that result in the horrible suffering of others makes perfect Darwinian sense. That a disciplined veteran would sacrifice himself for a mission or to save the life of his men can still be justified in the great scales of science. That people would sometimes do good for strangers, heroic acts done in direct disregard for one's own life, *that* makes a person begin to walk along the path of the *Uncanny*. We begin to hear God's whispers. The presence is felt of an *Other*.

In the biological hierarchy of our species, the woman of childbearing age and the child himself are the pointy top of the pyramid. Our tribal resources should go to keeping them alive, even if men must be traded for them. That a man would not hurt or kill a member of his own tribe makes perfect sense. The sentence that he might hear in his mind, "I should not kill that man." Or "I should protect that child even if it hurts me." can be easily justified by science. We evolved for tribal society. That he might take risks to save *another tribe's* child or woman could make sense as well, given that once rescued, they could be added to his own tribe, by their will or against it. Were he to risk his life for *another man*, he would be gambling one loyal tribesman for another who may or may not become a member. Williams didn't die because he saved a woman of childbearing age. He was in the water, passing the rope to men who *by definition* were less biologically fit than he was. Not only was he the one most coherent in the water, despite the brain addling effect of hypothermia, he was more alert than complete strangers, bold and decisive enough to organize a rescue effort, and physically fit enough to carry out that rescue. By any standard of Darwinism, he should have been the second one out, if not the first. Nature selected against lethargy and bad genes. Williams, in the words of the *Time Magazine* essay written by Roger Rosenblatt in his honor,

So the man in the water had his own natural powers. He could not make ice storms, or freeze the water until it froze the blood. But he could hand life over to a stranger, and that is a power of nature too. The man in the water pitted himself against an implacable, impersonal enemy; he fought it with charity; and he held it to a standoff. He was the best we can do.

What Williams did that day was that for which we have no answer. To spend too much time thinking on *why* a man of means, with bright hopes for a future with a woman he loved, would sacrifice himself for total strangers starts us down an uncomfortable road of why we idolize such insane actions. It leaves us with more *why* questions than we can answer. If not the workings of God, than how can man reconcile desires in himself for a righteousness that benefits the weak at the expense of the strong?

And what of "The Righteous Among the Nations," the Israeli designation and remembrance of those who risked their lives, and often sacrificed them, to save Jews that they often didn't know. What gets me about many of these stories is that the rescuers often didn't even know that the Holocaust was happening: they risked everything to save neighbors and strangers, alike, from what *might be occurring*. And the Holocaust is as good a place to continue this discussion as any other. That many people died needlessly at the hands of monsters is a fact the world is fond of reminding us. But it is "baked into the cake," if you will. That evil things happen, we all acknowledge. They are easily explained by the simple fact that a tribe will value its own members over others. In 1939, Hitler was allowed by the advent of industry to impose murder and sorrow on an industrial scale. But even during the Holocaust you saw, through the trees as it were, shining glimpses of *the Other*. In his classic *Man's Search for Meaning*, Victor Frankl, a psychologist and Holocaust survivor, famously wrote, "The best of us did not survive." A chillingly humble phrase, to be sure, but it is easy to miss the more subtle message: there were those who were confronted with great evil and responded with righteousness. They were the best; they could have survived by compromising who they were, but choose not to. They turned Darwin's savage equation on its head. They proved the meek could inherit the earth.

And here we find Christ. Not the smiley, white guy with a beard who holds a lamb, not the imperial king in a painting who has a halo around his head: we find the one who "turns the world upside down." We find the One who truly does roar like a lion. Here we find the one religion that says the last shall be first. The one faith that acknowledges that all the world agrees on basic standards of right and wrong and breaks those standards every day. A world where we are keenly aware that very little is as it should be. We spend nearly our entire lives trying to avoid what we all know to be inevitable, and occasionally commit great acts of wickedness, treachery, and violence to suspend our sentence. And yet, in that mixture, there are those who gladly accept death, who do good directly to their own harm, for people they should care nothing about. And this is where we are truly confronted. According to Darwin we should think those people deranged—instead we make them our heroes. Where Christ no longer becomes the One in the stories but the One in others. When we tell those who do not believe this they ask for facts. They require proof. They say, "Well if He is real, why doesn't He just appear right here and all shall believe." To which we reply first to our sorrow and then to our delight that He did. They said the same thing to Him then, and His counterargument was self-sacrifice. He loved us, we rejected Him. He performed miracles, we hated Him. He saved our lives, we crucified Him. It was the most unrighteous of all rescue missions: those saved were not together worthy of their savior.

Chapter 15

The Creation of Harris

Biology will never help us decide right and wrong; science makes for a poor God.

And so we arrive at the divide in the issue. Our party that has walked this path together now sees a fork in the road. I'm no Unitarian, and I do not believe Jesus was a great teacher. He was either exactly who He claims to be, or a monster. The Law and the Prophets were either really for our own good (and others, to be protected from us) or they were an especially abominable tyranny. The path of Who or What made the world now ends at a fork. The one, I believe, leads to life. The other has no rational and logical end but in *Nihilism*. Put simply, it is the belief that life is inherently meaningless. Your brain is just a series of chemical and electrical firings that affect other life forms, and in a billion years nothing anyone ever did will matter at all.

Fyodor Dostoyevsky was born in the very twilight of the Enlightenment movement and was raised on the ideas of its greatest thinkers. Raised in a family of humble means in Russia, he became a profoundly religious Orthodox Catholic. He was sent to a military college where he became known for his evangelical approach to the faith and appeals to the importance of love for one another being the greatest Christian imperative. His deep faith led him on a trek of questioning the pursuits of men and the place of government. Specifically, he became philosophical about the rise in technology and its ability to create incredible wealth and unimaginable suffering, and began to theorize about the place of a creator and objective truth in societies where the end of material need was in sight. In his classic, The Brothers Karamazov, Dostoyevsky has a poem where The Grand Inquisitor, embodying all that is wrong with

Catholicism in particular and tyrants in general, is torturing Jesus:

> But dost Thou [Jesus] know that for the sake of that earthly bread the spirit of the earth will rise up against Thee and will strive with Thee and overcome Thee, and all will follow him, crying, "Who can compare with this beast? He has given us fire from heaven!" Dost Thou know that the ages will pass, and humanity will proclaim by the lips of their sages that there is no crime, and therefore no sin; there is only hunger? "Feed men, and then ask of them virtue!"

Science is a poor God. If you remove the transcendent morality, that man can rise above the firing of chemicals in his brain and the mere quest to survive as the fittest, you remove the purpose that inspired the American Experiment. You remove our high desires of freedom and equality, and not just the satisfaction of our stomachs or the release of dopamine. In the words of Dostoyevsky:

> "The world says: "You have needs -- satisfy them. You have as much right as the rich and the mighty. Don't hesitate to satisfy your needs; indeed, expand your needs and demand more." This is the worldly doctrine of today. And they believe that this is freedom. The result for the rich is isolation and suicide, for the poor, envy and murder."

If you are, as Bertrand Russell claims, "a randomized collocation of cells" than, like it or not, your life has no more purpose than a giraffe or a centipede or a daffodil. You are just one more species striving to pass its genes on. The Atheist thinkers claim in different ways that we have meaning because it is entirely possible (if not likely) that the only sentient life in the universe is man, and our lives are valuable because of it. But that is only to say the human race has more value and meaning, not the individual. The trend of Atheism is towards the survival of the fittest. That Hitler and the Nazis embraced a eugenics movement started by Karen Sanger in the United States is no mystery to me. I 'm only startled, based on the evolutionary evidence, that it didn't catch on more broadly. Atheists will divert away from the belief in Nihilism—that everything is essentially meaningless and by extension life is really only about pleasure, suffering, and power—but only because they are being influenced by a Judeo-Christian ethic of the inherent value of individual human life. If we're just a "collocation of atoms," than why does anything have meaning? The truth, of course, is that it wouldn't.

I came across this very conversation in an opinion piece on www.richarddawkins.com (a pretty gangster move to name your own website after you, to be sure). The top comment said that the writer should not focus on all the "billions of years," and just try to effect change today. The post finished by writing, "nihilist in denial". And that is what New Atheism is, of course. Many thinkers have obviously run into this problem with morality. In his book, *The End of Faith*,

Sam Harris intuits this obvious problem with the Atheist worldview:

> Many people appear to believe that ethical truths are culturally contingent in a way that scientific truths are not. Indeed, this loss of purchase upon ethical *truth* seems to be one of the principal short-comings of secularism. The problem is that once we abandon our belief in a rule-making God, the question of *why* a given action is good or bad becomes a matter of debate. And a sentiment like "murder is wrong," while being uncontroversial in most circles, has never seemed anchored to the facts of this world in a way that the statements about planets or molecules appear to be. The problem, in philosophical terms, has been one of characterizing just what sort of "facts" our moral intuitions can be said to track-if, indeed, they track anything of the sort.

It is here that Harris gets dangerously close to agreeing with me. He comes so near to poking a curious eye through a gaping hole in the Atheist argument, it's infuriating. Almost as if, in that great painting by Michelangelo, *The Creation of Adam*, Harris nearly touches the outstretched and straining hand of God. Instead:

> A rational approach to ethics becomes possible once we realize that questions of right and wrong are really questions about the happiness and suffering of sentient creatures. If we are in a position to affect the happiness or suffering of others, we have ethical responsibilities towards them-and many of these responsibilities are so grave as to become matters of civil and moral law…
> …Our ethical intuitions must have their precursors in the natural world, for while nature is indeed red in tooth and claw, it is not merely so. Even monkeys will undergo extraordinary privations to

avoid causing harm to another member of their species. Concern of others was not the invention of any prophet.

I will work my way backwards, after simply mentioning that this is not out of context and Harris's genuinely well-written book devolves into propaganda in the chapter, *A Science of Good and Evil*. His source citation about monkeys is found in *The Swappable Minds*, an essay written by Dr. Marc Hauser. As I said earlier, someone can say a true thing and still be wrong about other things. Karl Marx likely had several good qualities, but to quote him positively on *Political Science* would obviously be a mistake, given his record. Dr. Hauser studied the social fabric of monkeys during his time at Harvard. Dr. Hauser also was found guilty of eight counts of scientific misconduct—including at least one confirmed case of falsifying statistical data. He was suspended and then resigned in disgrace. That Harris cited a fraudulent study is somewhat excusable; he does have a bibliography to fill (one that I personally think is stuffed for credibility's sake), but that he perhaps didn't check every source is understandable. I knew that this citation was likely fraudulent though, because of the sheer ridiculousness of the claim. Monkeys, like all animals, don't give a hoot what they do to other animals, and certainly exhibit a pattern of wanton violence against their own species. What we might anthropomorphically call, "rape" and "murder" is common to every higher-intellect animal, as Adam Rutherford points out in an article for *The Atlantic*:

> Infanticide is another unpleasant behavior seen in dolphins. It often gets translated into murder in the popular press, but it should be noted that in plenty of other organisms, both males and females kill the young of others within their own species as a reproductive strategy.
>
> A female lion lactates for more than a year when she is nursing cubs, and during this time won't breed. Males acting alone or sometimes in packs will kill her young in order to bring her back

to being fertile, so they can then sire a pride. Mother-and-daughter teams of chimps in Tanzania have been seen killing and eating the babies of other parents for reasons that are not clear.

In an article for BBC Earth, Michael Marshall writes about how Red Colobus monkeys were nearly made extinct by chimpanzees. Chimps are known to not just hunt other monkeys but, according to Mindy Weisberger of livescience.com, are known to enjoy eating the brains of other chimps. So no, Mr. Harris, I'm pretty sure that "biology class" isn't where we should go for the construction of our moral institutions. But he isn't the first to suggest that we find our morals in the natural world. *Social Darwinism* is the belief that "survival of the fittest" is the natural order, and that any delineation from that natural order (minority rights, accessibility for the disabled, etc.), posed an existential threat to humanity by allowing inferior genes to continue. If you think your brain has no inherent divinity or that there is no objective truth, what you are really saying is that an evil thing done to you or your group is only "evil" in that you don't happen to like it.

What about if someone steals your bike and they go to jail? What about *their* unhappiness and lack of pleasure? Is there a way to quantify suffering, so that they can be doled out a proportional discomfort in response to what they did to you. An "eye for an eye" if you will? What about breaking up with someone? I had a girlfriend and realized that we didn't have a future together. I broke it off and she was distraught. What about *her* happiness? Should I be punished for what I did? I acted in a way that I thought was best for both of us in the long term but she certainly didn't agree. To say that we should *always* attempt to avoid suffering in others is obviously non-sense.

In combat, for example, two soldiers fighting against each other are both trying to live. They don't give a second thought to the agony their enemy will experience in death, as long as they are the one to survive. Only someone who has never risked their life in the insanity of combat could say that a flamethrower is "unfair." And suppose we're talking about the Allies vs. the Nazis? Should the American G.I. be able to inflict *any* amount of suffering on the Wehrmacht infantryman because, in hindsight, we were certainly justified? What "level" of suffering is it ok to inflict? I would love an answer from Harris or Dawkins, but I know the truth is that we will never get one. Because Biology can't give you answers on ethics, any more than math can correct the grammar of this sentence. Science makes for a poor God.

In the case of Natural Law, we could argue that there was nothing inherently "evil" about the Holocaust, as good and evil are relative entities. Morality loses its meaning entirely. If anything, the Social Darwinism that Christopher Hitchens decried in his book, *the portable Atheist*, should be considered a net positive. If *good* only means that which aids the progress of humanity (as both he and Sam Harris are so fond of pointing out), then, far from being evil, Eugenics and Social Darwinism are both *necessary* and *good*. If only the collective progress matters, then the individual will always be subjugated to the good of the group.

Such Darwinian belief systems led to the forced sterilization of some 64,000 immigrants, blacks, single mothers, and mentally handicapped in the United States from 1919 until well into the 1940's. And, as Eugenics can't even be discussed without referring to Germany's Third Reich, remember that, within just twelve, short years, Eugenics metastasized from sterilization of the "damaged" members of the Reich, to the elimination of "Useless Eaters" (the elderly, mentally handicapped, physically deformed, etc.), to the proposed destruction of an entire race. Without a belief that humans are *inherently* valuable as *individuals*, and *uniquely important* to the community, you are left with science. The worship of science, in fact, in less than a century, has given the world a higher body count than *every* recorded war *combined*.

This is the main issue with Atheism. It is why I wrote this book. In too many ways Atheism proposes a breakdown in *objective truth*. Put most simply, science was never meant to be our God.

Chapter 16

The Carpenter Who Builds a Bridge

We must learn how to meet people where they are.

It really is a shock that I ever came to work for an organization like Younglife. In many ways I raised myself, and by High School I was so used to not trusting people that I had a hard time making close friendships. Like a group of coal miners, the kids and volunteers took turns chipping away at my defenses. One thing led to another, and over the years I began to make friends in the popular youth ministry. Maybe it was the fact that it didn't bother with sacraments and theology. The volunteers, mostly 20-somethings, seemed to only concern themselves with making a difference in the lives of adolescents, and specifically, my life.

Later, as a sophomore at a military college, I had become a volunteer "leader" in Younglife for a few reasons, but my primary motivation was in realizing that while I couldn't go back in time and solve the problems of my childhood, I could affect change in the lives of young men struggling with similar issues in the same, crumbling social structure I had once languished in. I was assigned (placed, in the parlance) at a private Catholic school. It didn't help us that the school thought Younglife was a protestant group of hipster, fifth columnists, attempting to subvert the parish. Nothing could be further from the truth, of course. I had a difficult time trying to meet high school students to begin the process of making real inroads with them. The mission was to become friends, and through those friendships, mentor and counsel the young people. If they wanted to hear about Jesus, great. But they knew we would be still be pals if they wanted to be Muslim, Hindu, or in the case of one of a dear friend of mine, an Atheist and Communist party member (he was an exchange student). It was the safety and trust of knowing that I "was in it for the long haul with them" that often encouraged these adolescents to listen to me long enough to hear what I had to say.

One day, my team leader suggested that I might try coaching. The school had a new rugby team and I had played in high school and during my freshman year of college (until permanent damage to my knees forced me to retire). I took my schedule apart hour by hour that following week. When I emailed the head coach, I exaggerated my skills so fiercely that it was mostly a lie. He asked me to come to the school, where I sat through the Catholic Church's intensive screening process—and then was given a whistle. I was the team's fitness coach, and all it cost me was spending all my money on gas and more time than I had on paper to give.

A few months in, and I was presented with a decision that likely changed my life. We played our first game on the other side of the city. Where I live there are multiple islands and peninsulas separated by bridges. The most distant bridge from me was where we had our game.

The boys won that first game on their way to a perfect season—and a state title. The mood that night was high, given the other team didn't score a single point. I was collecting my things, ready to make the long drive back across multiple bridges, to my barracks after a long day. It was late already on a Friday night, and I had to be up early the next day, after a particularly long week. Even worse—not getting enough rest on the weekend in that environment meant one was sentenced to begin another hard week at 5am on Monday, while already being tired.

One of the players was talking to the others, inviting everyone to Waffle house after the game. The famous Southern "We'll-serve-y'all-anytime-of-day-no-questions-asked-come-sober-or-not" staple was two peninsulas and an island from where we were but it was back where the boys lived, but not where my college was. They looked over and realized I could hear. One of them said,

"Oh uh, Coach Matt, you can come too. It's the Waffle House over on Long Point Road."

I replied that I probably could, but that I would have to check. I woke up at 5am that morning. I was one tired boy. It's hard to really convey how little I wanted to drive all the way out to a Waffle House at that hour. I knew the players, but not well, and even if the goal was to get to know them better and one day be able to just enjoy a regular, late-night breakfast with them, I wasn't sure that it was the best time just yet.

I ended up driving back after most of them began to leave and even though I felt like I should have gone with them, the excuses I made to myself seemed compelling. As I rolled through the front gate of the school, the final stop sign lit up in my head lights. I just had to make one more left turn, park, and then I could go to bed finally.

At the stop sign I took a right turn—and I dared. I dared to break the convention of "keeping the coaches and players apart." I dared to lose sleep and gamble a whole week on what would likely be one of the more awkward meals of my life. I dared to be a positive male role model. I dared to know these guys.

Thirty minutes (and three bridges) later, my car pulled up to the Waffle House. My heart beat was getting faster. It wasn't just feeling awkward, I actually didn't belong in that social situation. The guys had not even gone inside yet—when one of them spied me inside my car while I parked at the end of the lot. He shouted, "Hey, its Coach Matt!"

And then, just like that, my life changed. Because in that instant they began chanting my name and laughing. What followed was the first of many trips to Waffle House. Overtime I became, in some cases, the only positive male influence in the lives of several of those young men. Over the course of the following year, I began eating dinner every Wednesday night at Zach's house, one of the guys there that night. I still go to dinner with him every week and we both refer to his mom's mom as "Grandma." I spent every birthday for the next three years with the boys, typically coaching. They are now some of my closest friends. I believe that they have given me more than I have ever given them—all because one night, I set the tone. I dared to know them.

I did it because the One I follow builds bridges, not walls.

The best possible thing that has happened to both the religious and the Atheist alike, is the age of information. The Gutenberg Press made it so that everyone could read the Bible for themselves, and in doing so, robbed the Catholic church of their monopoly on information. Social media and the internet will do the same for those oppressed by the ruling ideas of their culture. It also represents an ideological "natural selection", which will select against weak ideas. Only the strong will survive.

However, that concept is near to being the direct opposite of those ideas espoused by thinkers like Sam Harris. The unfortunate story that history is far too fond of telling is the propensity for a tribe, especially a tribe of the religious variety, to press their advantage and crush the competition. When an ideology has no ethical mooring—except to defeat an oppressor, you get carnage. *The End of Faith* by Sam Harris is probably as well-intentioned as Karl Marx's *The Communist Manifesto*. And if Harris and New Atheism succeed in creating a political movement that pits the antagonism of religion against theism, and if some revolutionary group then latches on to the insanity of the ideas he espouses, I'm not being dramatic when I say the death toll will be counted in ten digits. Take the third page of his book,

> But technology has a way of creating fresh moral imperatives. Our technological advances in the art of war have finally rendered our religious differences- and hence our religious *beliefs*- antithetical to our survival. We can no longer ignore the fact that billions of our neighbors believe in the metaphysics of martyrdom, or in the literal truth of the Book of Revelation, or any of the other fantastical notions that have lurked in the minds of the

faithful for millennia- because your neighbors are now armed with chemical, biological, and nuclear weapons. There is no doubt that these developments mark the terminal phase of our credulity. Words like "God" and "Allah" must go the way of "Apollo" and "Baal", or they will unmake our world.

Paradoxically, Harris takes part in a conversation now referred to (and which is available in book form) as the *Four Horsemen Discussion*, just after the publishing of his bestseller, *The End of Faith*. Participating in the conversation are Richard Dawkins, Christopher Hitchens, and Daniel Dennett—who gives a vision for the future that I whole-heartedly agree with:

> When people tell me I'm being rude and vicious and terribly aggressive I say, "If I were saying these things about the pharmaceutical industry or the oil [industry] interests, would it be rude?

> Would it be off limits? No. I want religion to be treated just the way we treat pharmaceuticals and the oil industry. I'm not against pharmaceutical companies, I'm against *some* of the things they do, but I just want to put religions on the same page with them.

Dawkins wonders aloud how religion got its "charmed [protective] status" before remarking on the hypersensitivity to offending people's religious beliefs, and the exceptions warranted by them. Both gentlemen have very valid points. Earlier in his remarks, Hitchens made a noteworthy exception that he certainly was opposed to other non-believers doing things that are obviously offensive (depicting lewd acts of the Virgin Mary, disrespecting or vandalizing religious institutions and art work, etc.) He simply didn't understand how it can be possible to allow religion to enjoy a status, whereby a person can't criticize a belief system that someone uses as an excuse to commit evil acts (an Islamic fundamentalist using the Koran to excuse terror attacks for example).

He also didn't understand how we can have a society that evolves and progresses wherein a majority of the country has their belief systems protected from rational debate by a minority of the country. It is even more noteworthy that Hitchens and Harris both agree that Atheists must "play ball" as it were, and strive to not attach emotional arguments to their ideas about the universe, lest they fall victim to the same outrage culture that religious fundamentalists can be guilty of. And they are, of course, exactly right. We will never begin the process of rebuilding our common dialogue if people feel compelled to run home the second their beliefs are seriously questioned.

In the last two sections I have quoted Harris in two different mediums—one, his book, and the other, a conversation that really created the New Atheist Movement. In the former, he takes a deeply antagonistic note, and in the latter, he almost borders on the conciliatory. To the first, I offered a grave warning; with the second, I find myself nearly agreeing.

So why cite both and not one or the other? Well, the answer is twofold.

Firstly, in line with the theme of this book (along with my personal code of ethics), I cited both because I wanted to give the best possible view and the fullest expression of the opposition regarding this book's thesis. It would be easy, if not altogether advisable, to paint Harris and company as a bunch of radicals bent on destroying this country. I refrain from doing this because, for one thing, I don't happen to believe it. For another, my country (And I wish that people in other countries couldn't, relate but I fear they can) is filled to the brim with that kind of negative, nonproductive dialogue. This book is not a political one. I don't write here about what will only be important for the next twenty years, but the next 20 thousand, 20 million, and so on and so on.

Second (and somewhat more to the point), I like to think that all of us can get worked up into a frenzy when we get our emotions caught in the chain of topics that deserve a more rational and nuanced approach. I notice that no matter how bizarre an opinion might seem to me, I find that if I listen to someone calmly describe their thoughts, I at least can often see their heart behind it. The vast breadth of human experience continues to amaze me, and through keeping an open mind, I find that I have become more entrenched in some beliefs, while much more relaxed in others. I've found that, over time, *my initial reaction to information is less extreme.*

For example, when I began the research for this book, I invested countless hours in learning the best arguments for Atheism, secularism, Communism, Marxism, and moral relativism. What I learned in my countless hours studying the works of numerous intellectuals, is that while I felt more confident in my opposition to many of the points they raised then when I began, I also discovered—to my surprise—that I would likely enjoy talking to them in person. The odd quote from Dawkins or Dennett, which would have riled me up when I was younger, merely elicited the desire to get the full context of the author's ideas, including the medium it was presented in—and how rehearsed their remark was. Perhaps this process has elicited a certain *humility* in me; a better understanding about how easily a flippant, offensive remark can be uttered. This is especially true when discussing those concepts/ideas that one fervently believes in.

A man by the name of Steve Chesney, the Regional Director of Younglife in Tennessee, walked around an empty camp with an interviewer. It was spring in North Georgia, so at Sharptop Cove, a beautiful summer camp nestled in the valley between several mountains, the birds can be heard singing in the distance, while Steve talks with the host.

> …kids today are also different. They've experienced an incredible loss of innocence—sending and receiving sexually explicit photos is the common language of the day. I think also, in recent years, in our culture and certainly for kids, that Christianity has become more and more irrelevant in their minds. Kids have become more skeptical and

cynical, its an age of doubt, and they come by it honestly. They can google up any question and have 100 answers in a nanosecond; *but who are you gonna trust?...* ...I think *authenticity* and *humility* go a *long* way. I don't think they're looking for another talking head who is arrogant, a know-it-all, judgmental, intolerant, gay-bashing. I mean certainly I think its our job to hold out the *truth*, with certainty, unwavering; but also to hold it gently and with compassion. And maybe even to start a sentence with, "And you know, I may be wrong, but here's what I believe and here's why." And allow kids to come to their own conclusions.

And that is near the heart of the issue. There is a distinct reason why this book stays so far away from politics: in ministry, I always work to leave all that nonsense at the door. It's always a philosophy of mine to, as the example was modeled to me, *meet people where they are*. I don't think anyone ever became a Christian in an environment where a wooden club was involved (at least not in any meaningful sense). I don't think we will ever narrow the growing divide if we don't meet each other where they are. Me and the guys are all bowling tomorrow night. Don't tell them that I would never go bowling for fun. I may hate it, but I love them. Me and a guy I worked with last fall are going fishing when it warms up. He's not sure what to believe about God; he also majored in history (we few must stick together), and a great fisherman. When we talk religion, as we do often, I guess I just find my sentences often starting, "Well I could be wrong, but here's what I think…"

Jesus had some strong suggestions about loving your enemies. He didn't call his political rivals a threat. He died for them. The Carpenter made bridges, not walls.

Chapter 17

The Closer You Go the More You Find

It's not that the unbeliever goes too far into the sciences, it's that they haven't gone far enough.

I once learned how to navigate in the woods. Not just compasses and rights and lefts, but the whole nine-yards: protractors, grid maps, azimuths, etc. It was outdoors and remote, so we carried our packs with us, as we tried to find our way to the stakes that had the set of coordinates on them we were looking for. Between most of the points were land features, known as the "draw". A draw is where a small stream, sometimes no more than a few inches thick, creates a dip in the land. These draws are swampy by nature, and consistent water supports dense vegetation. That vegetation is often thorny vines, and, you guessed it, more often than not, I had to cross these draws to get to my point.

And yet, despite all of this, I often walked into my points, at nearly the exact distance of my "pace count" (you count every time your left foot hits the ground—for me 55 left foot steps is approximately 100m. It can be strangely accurate if you develop a sense for estimating the change in elevation and drift). I became so good at it people would come to me for advice. I beat guys that taught it to other students. Heck, I beat guys who had to do it to survive in Afghanistan. No one could believe it, me least of all.

If I ever made a mistake, it was that I would not trust my pace count. I would get off early, I would take the wrong hill or stream or spur, and end up way off-point (and cost myself tons of time and worse, extra miles under that pack).

If there is one logical fallacy that, in my opinion, stands heads and shoulders above the rest, it is that of the intellectual who, when seeing that evidence credits them, gets off the path and says, "Aha! It is as I always expected." They misinterpret the hill, and get hopelessly entrenched in the draw

You know what the worst thing about a draw is? It's not the time you waste or the slashes you get. It is the near-impossible and humiliating choice of either reversing out and getting a known compass bearing—or continuing to turn and slash and fight and curse and to inevitably be more lost than when you started.

We get so fixed on the target, or goal, or political policy, or friendship, or idea, that we lose focus of the broader picture. And sometimes the closer we get, the more dangerous the temptation becomes to cut corners. Once, a friend of mine found his way into a draw so large it was actually named—and had a running river in the middle of it. He thought he was just in a lowland swamp, as there were plenty of them, and none were especially hazardous—right up until the moment he stepped deeper into the current and felt only moving water underneath him. The inky blackness swallowed him—country accent and all—and away he went, hopelessly downstream. He, very literally, missed the point, and we all relentlessly made fun of him. It was the stuff that legendary nicknames are made of.

And I think that's the long and the short of it. Atheists are not evil, mustachioed villains, who go too far in the sciences. They love science and logic and physics and reason- traits they get from the Great Scientist Himself. The problem is, they never go far enough. They lose perspective, and end up missing the point. Sometimes we take steps too far into the deep because our goal seems so close—and then we get carried away. I can only imagine the wonder of building theoretical models of String Theory, Gravity, and The Big Bang; I fear, however, that being so close to such amazing mysteries might tempt the scientist to lose perspective on the greater picture. It's well that you study our origins as a species and our planets and our solar system, but lose sight of the fact that it all had to begin, and more importantly: begin from What or Whom, and you have thoroughly missed the point.

I suppose that trust is the trait that made me good at finding my way. I put trust in my pace count and the azimuth that I shot. It sometimes meant fighting my way through something thorny and unpleasant. The temptation was always there to give up, when buried to my waist in muck, and cut a million times by the undergrowth, but I would keep going. Like magic, eventually the weeds parted, and a yellow stake with my next set of coordinates was there to greet me.

Perhaps the answer to our doubts in troubling and confusing times is not to surrender but to persevere. In the words of Tim Keller, "Is it better to have weak faith in a strong branch or strong faith in a weak branch?" We doubt, we're challenged, and we're wrong countless times; we repent. We're knocked down; we persevere. Christianity is the religion of the always forgiven, not the always right.

And speaking of Christianity, what of their Christ? That one Man could change the world so soundly, so radically, so completely, in a way that the likes of Xerxes, Nebuchadnezzar, and Napoleon could never dream of? The one born in the last century BC who would alter the earth so fundamentally, would not be Julius Caesar, the one who slayed the Roman Republic and began the Roman Empire, but some *lunatic* born in a Jewish backwater, murdered by his own people?

Charles Colson was charged and sent to prison in 1973 for his involvement in the Watergate scandal which led to President Richard Nixon's impeachment. He soon after became a Christian and then pleaded guilty to his involvement—against the advice of everyone from his family to his lawyer. In prison he began a ministry that now includes 110 countries and has dedicated his life to advocating for those incarcerated and those newly released. What convinced him that this Jesus was Who He says He was? A combination of C.S. Lewis's *Mere Christianity*, and a thought that occurred to him in jail:

> I know the resurrection is a fact, and Watergate proved it to me. How? Because 12 men testified they had seen Jesus raised from the dead, then they

proclaimed that truth for 40 years, never once denying it. Every one was beaten, tortured, stoned and put in prison. They would not have endured that if it weren't true. Watergate embroiled 12 of the most powerful men in the world-and they couldn't keep a lie for three weeks. You're telling me 12 apostles could keep a lie for 40 years? Absolutely impossible.

That his followers, eleven fisherman and tax collectors, uneducated and untrained, could keep a story straight not for three weeks but for forty years? A story which the historian Theophilus would corroborate decades after the fact—nearly word for word—in his *The Gospel of Luke*. Scoffers will tell you the church founders wanted power and "made it all up" in a bid for influence and control and probably money. Fascinating, though, that the egotist and tyrant who would do such a thing would then make himself and his ilk out to be so untrustworthy and foolish as the gospel accounts portray their alleged authors. How is it that these followers could convince thousands of Jews (who were more educated by and large than nearly any other people group until the 20th century) that their Messiah was not only *nothing* like what they expected (indeed He came as the Lamb, not the Lion) but that He had already arrived and they had crucified Him? Remember that people in those days may have been uneducated, but that doesn't make them stupid. The great conquerors of the world today are largely forgotten, their realms and dominions evaporated back, in nearly every case, into the rough approximations of the ethno-cultural groups they originated from. Great orators and authors are remembered, but what of it? How many missionaries bring the works of Seneca and Shakespeare to hostile cultures knowing the mission to be suicide?

And here we are with the earth upon Scoot's grave barely settled. But what of his killers? What justice shall be given for the poor little guy, or what meaning can we take from his end? As we have seen he is dead but should not be. He was not wrongly slain, but the method was wrongly scientific. That a parasite should have such a capability is impossible; but there he lays. And now that I mention it, we start to see other flaws in the science, leading like a trail of bread crumbs to a guilty child. The field mice were killed by the owl but his wings don't make sense. They flap and he flies, yes, but in what generation did an ancestor first think to do so? And, more pointedly, how?

I suppose if we're going back we might as well go all the way. Strange so many feel comfortable with a belief in a world that did not technically begin. We have much now, but from where did it come? If the science is air tight, if the convictions around Atheism so steadfast, why then are the arguments surrounding it always so emotional? Why care so much about nothing? I wonder how many other major scientific systems that have infinitely important religious, philosophical, and mental ramifications are created "…without any working theory as to how it all began."?

Certainly, Christians don't agree on everything, no one, in any time or place, has. It makes more sense to me that issues of such importance are hotly debated, it would be strange if they were less. That many believers will scoff at the claims of an old earth, I have no doubt; I care, and doubtless they care as well, only that one day we will rejoice together when that great day comes and every promise that was hinted at on earth: a mother looking at her child for the first time, a man on the front saving his friend from the enemy, a poor man giving a poorer man a few dollars; will all be fulfilled for that which they really were. The place where all those feelings of joy, and beauty, and sacrifice, and belonging, and trust, and satisfaction actually came from. You will find your rest in Him because you were made for Him.

And we are his greatest creation. The masterpiece of the Artist. He creates. He is Creation. The Enemy destroys, breaks, and divides, but He puts back together. The animal world is one that takes from others in order to prolong itself, before its inevitable demise. The caterpillar eats the leaf of the Oak tree, which was perfectly content to stay as it was. Even in all its splendor, the butterfly will soon be nothing. The beaver creates dams which flood lowlands. Whatever nature makes, it makes at the cost of something else, and will never make that which is more complicated than itself. But in man, we find something Higher. We find the creature that can make things that are beautiful *for the purpose of their beauty*. Be it cave painting or *Pieta*, we have the creature who can create with unfathomable skill (for what purpose science will certainly never find a real answer). That creature can also destroy with apocalyptic violence. And in that dichotomy, we again see the intentions of the Creator, the story of Creation (that man would be raised from his origins as a beast into a kind of divinity), and the great pendulum of free will. Atheism is not new. We have always had to decide between prostration before Infinite Himself or the rejection of an *Other* for the worship of ourselves and what we can do and see.

How does everything come from nothing? How does everything tend—quite fortunately in our case—towards structure, order, and life, when by law it must tend towards disorder? Why does this law of deterioration only seem to apply conveniently after the settling of our world and all worlds? But of course this is all nonsense, and I think everyone, be they Richard Dawkins, your second cousin, or the Pope knows that there really is a Creator. And that everything is from Him. That He was, and is, and will be. That He is the Artist and we are the creation. We will create ourselves, indeed in His footsteps and to His delight, but never will we make that which is more sophisticated than ourselves. And He never created that which is greater than Him. Nothing can be more than the sum of its parts. Rightly it was said more than once, as the first answer and the last which will mean anything long before and long after all else has become meaningless: I AM. And so He Is. He is the End of our biology, physics, and chemistry, because Science is the only time that the art ever gets to admire the Artist.

Acknowledgements

I would first like to thank the young men who are the reason this book was written. My "muses" were 40-50 guys who I have had the profound pleasure to mentor and coach. It was long nights listening to family problems, car rides to adventures that nearly killed us, laughing in public places so hard we cried, and countless early mornings and late nights that inspired a book where I summed up my entire spiel of years in Younglife and youth ministry: your life *matters* because you were *made*. This book would not and could not have been written without you. If you are reading this book and wonder if I still remember you, I promise, I do. I specifically would like to thank thirteen young men: Zach, Jack, CJ, Jorden, B Dog, Jeremy, Josh, Preston, Trevor, Cole, Pomme, Harry, and Tobin. It was a hard season for me leaving Active Duty (it's a long story) and you guys welcomed me back to Charleston with open arms. I was your Younglife leader in high school, now you all are some of my best friends. This book could not have happened without you; you are the reason I wrote it.

I would like to thank my biggest fan, my mom, who has a perfect record of complimenting my work before softly criticizing it. Never, from my first words to this book has she been anything less than supportive. Her tireless effort and enthusiasm made this book possible. I would like to thank my father for his overwhelming support for my career and harsh criticism of the first draft of this book. I say that with a smile on my face. He hated the first copy and told me not to risk my Younglife career. Conversations with him molded the third chapter. In fact, his criticism led to me taking this manuscript back to the drawing board. What I had was not what I needed. He has my eternal thanks for his input. If you don't have people willing to tell you something sucks, then you don't have family.

I thank several teachers throughout my childhood and all of my professors from the History department. Specifically, I thank Mrs. Gurley from Aiken Prep. She assigned risky books that were fun for boys to read. She encouraged young men to do things besides play sports and be good at math. I wrote stories in her class that were the building blocks of my career. If my friends are *why* I wrote this book, my parents and teachers are *how* I wrote it. Dr.'s Preston and Sinisi at The Citadel were the first been-there-done-that authors to tell me I could make it in this business. It meant an enormous amount considering how often they told us not to consider being full-time authors.

If my "guys" were *why* I wrote this book, my family and teachers are *how*.

To all my friends, colleagues, and mentors who, for free, suffered through early drafts of this work to provide the feedback I so dearly needed: a million thanks. I don't deserve you all.

I thank all my classmates at The Citadel; they always made fun, but they never stopped believing in me. On our first day the cadre told us to look around the room: we were looking at future groomsmen at our weddings. It was certainly true. I will die not having missed that place a day in my life, but I miss my brothers terribly.

I thank my church, Grace City, for their open doors and open hearts. I finally found somewhere to land in Charleston after so much turmoil and wandering. I have found a family there that is married to the Gospel and on fire for the justice of Christ. I couldn't be more satisfied.

Finally, I thank several mentors of mine—all in Younglife (I'm sorry mom, I know you're tired of all the Younglife shout-outs). Drew Dempsey, my role model and leader when I was an awkward high school kid with few friends and fewer social skills. There's a reason so many of your guys became leaders when we went to college. Dennis Fuller, the first man to ever stop me and really try to understand the pain I had kept bottled up so far away from my conscious self. I'm a work in progress and I couldn't have done it without you. Neil

Gardner, my Area Director and friend throughout years of adventures and misadventures. You took a flyer on me three times in my life and it changed who I have become.

Annotated Bibliography

"Oh. Hey there... uh, can I help you?"

"What's that? You want to read the *annotated bibliography*? You know that's where I explain every source with a sentence or two? Sort of like in school where your teachers didn't trust you did your work, so you had to prove it?'

"Oh. You did know that? Well, I guess it makes sense. Huh. Ya, well knock yourself out. I'm sorry I don't get many visitors 'round these parts. You need help with that stack? No, of course not. Sorry, I guess I'll just wait over there. Let me know if you need anything."

A few thoughts on my research method. It's important to know that there are relatively few *book sources* on Atheism whereas I have cited many Atheist websites and online authors. The reason for this, to reiterate, is that there are many books on the topic but few have popular support from what appears to be a majority of those who identify as Atheists. I have included book sources based on whether or not the book informed my study of the topic *or* my technical writing of it. For example, I'm a huge fan of *The Weight of Glory* and although I did not directly use the work within the text, I was deeply interested in Lewis's argument structure and narrative flow. I cited three works by him in this Bibliography because those three stand out for their help in this work (as opposed to his whole library, which I have read and recommend the contents of over anything I will ever write).

It is worth stating up front that my inclusion of these authors and compliments towards them say nothing about my personal beliefs other than that I can disagree with someone and still be polite (sometimes). An educated guess might accurately pin some of my political beliefs, but I do my best to keep them from the preceding book or this following bibliography. Finally, if you trust nothing else that I say, know that this bibliography is at least honest to the best of my knowledge. It is not packed, in my section on Scholarly Articles I list all names given in the paper to give credit to everyone to whom it is due. You will notice this creates the effect of greatly lengthening the bibliography; I only seek to honor the authors; if I could make this shorter I would. In further editions I will gladly update the printed version of this book as well as immediate updates to the e-book if there are any discrepancies that come to light in the future.

Books

Carroll, Sean. *From Eternity to Here: The Quest for the Ultimate Theory of Time.* (New York: Penguin, 2010). A fun book on the ideas of time by Carrol. I certainly don't agree with all of his conclusions but his work was helpful in my book.

Colson, Charles. *Born Again*. (Old Tappan: Chosen Books, 1976).

A beautiful book from a man involved with the "Watergate" Scandal. A triumph of the Gospel in the life of a man. I highly recommend.

Dawkins, Richard. *Climbing Mount Improbable*. (New York: Norton, 1996).

Dawkins, Richard. *The God Delusion*. (Bantam Press, 2006).

Dawkins, Richard. *The Selfish Gene*. (Oxford: Oxford University Press, 1976).

Dawkins wrote *The God Delusion*, which was built on the foundation he had laid with earlier books such as *Climbing Mount Improbable* and *The Selfish Gene*. For all our disagreements they are witty books and he makes excellent arguments. Unsurprisingly I referred to *The God Delusion* the most of any of his books, and I would suggest the same to anyone new to the topic. It is, in my opinion, the gold standard of the New Atheist movement.

Dennett, Daniel. *Darwin's Dangerous Idea*. (New York: Simon and Schuster) 1995).

A deeply technical book, which hides chunks of solid arguments and ideas within deep jungles of superfluous content. Among my more serious criticisms I decry the formatting, font, and overall length. Brevity is the foundation of wit, Mr. Dennett.

Dostoyevsky, Fyodor. The Brothers Karamazov. (New York: Random House, 1975).
Not only have I not read the entire book but I needed a later source in order to crack into its profound depth. On my reading list for sure, but right now it was helpful near the end of my research.

God. *The Bible*. (Jerusalem: Multiple Publishers, Debatable).

Harris, Sam. *The End of Faith*. (New York: Norton, 2004).

The great alarmist of the *Four Horseman*, I can forgive Harris for his Chicken Little routine in the 2004 book being written as it was in an era of fear for those seeing fundamentalist monsters in every shadow. It is also the most compelling of the books, and I have no doubt that were I still an unbeliever this book would have evangelized me towards the Atheist cause.

Hitchens, Christopher. *god is Not Great*. (New York: Hachette Book Group USA, 2007).

Hitchens, Christopher. *the Portable Atheist*. (New York: Da Capo Press, 2007). Hitchens, professional career notwithstanding, seems like the most reasonable of the *Four Horsemen* and I deeply regret we could not share a drink and talk before his untimely passing. I remind myself that God loves those who change their mind, even at the last moment. No one deserves the gift of Grace.

Jones, Malcolm. *Dostoevsky and the Dynamics of Religious Experience* (London, Anthem Press, 2005).

> See above note on Dostoyevsky.

Junger, Sebastian. *Tribe: On Homecoming and Belonging.* (New York: Twelve, 2016).

> One of my all-time favorite books. I regret only that you read this one before that. Without a doubt one of the best new books on belonging and community from a cliché liberal-writer-in-a-warzone who has at the same time been-there-done-that. A fantastic and short read. Dennett should take notes.

Lewis, C. S. *Mere Christianity.* (London: Bles, 1952).

Lewis, C. S. *The Problem of Pain.* (New York: Macmillan, 1944).

Lewis, C S. *The Weight of Glory.* (London: Society for Promoting Christian Knowledge, 1942).

> Lewis is, without a doubt in my mind, the foremost Christian apologist of the industrial to

modern age. Many treat his ideas as near apostolic in value. It is well earned. He is high on my list of those I wish to speak with in Heaven, it will be worth the wait in line.

Lukianoff, Greg. Haidt, Jonathan. *The Coddling of the American Mind: How Good Intentions and Bad Ideas are Setting Up a Generation for Failure.* (London: Penguin, 2018).

A fantastic book whose only flaw is that the audio version is about three hours too long, but forgivable given the clinical nature of the two authors, who no doubt seek a perfectly air tight work with no stone left unturned. The work was a great help in framing ideas I had about argumentative codling.

Pressfield, Steven. *The War of Art: Winning the Inner Creative Battle.* (New York: Rugged Land, 2002).

Pressfield, Steven. 2012. *Turning Pro: Tap Your Inner Power and Create Your Life's Work.* (New York: Black Irish Entertainment, 2012).

Fantastic and short books for any aspiring writer. Both were instrumental to my process of assembling my book.

Revelli, Carlo. *The Order of Time*. (London: Penguin Books, 2018).

> I have never read something that was so beyond my cognitive ability to understand. I also, for the record, disagree with many of the author's conclusions, but I have no evidence with which to firmly argue, as in many of those arguments Revelli has no solid evidence with which to argue from. That is not a criticism, on account that his work is mostly in the abstract and hypothetical realm of physics. A phenomenal book that will likely cover all of your time questions, or at least give you *an* answer.

Sagan, Carl. *The Demon-Haunted World: Science as a Candle in the Dark*. (New York: Random House, 1995).

> Sagan has some good points, but I fault the book for its sensationalism. It was useful for my

research and "Sagan's Dragon" is a fun thought experiment if nothing else.

Shapiro, Ben. 2019. *The Right Side of History: How Reason and Moral Purpose Made the West Great.* (New York: Harper, 2019).

Shapiro wrote an excellent counter argument to Leftist ideology in the 20th and 21st centuries. Particularly his writings on Dostoyevsky were helpful to me as I was trying to crack into the daunting world of Enlightenment and Post-Enlightenment thinkers.

Willink, Jocko. *Discipline Equals Freedom.* (New York: St. Martin's Press, 2017).

The best non-Christian book on my list of "Must Reads". It will change your life. Read it, study it, memorize it; tattoo "Good" across your forearm. Whatever it takes.

Television Interviews

Dawkins, Richard. "Darwin and the (im)Possible Evolution of the Eye." Interview by Howard Conder. Revelation TV.

> Dawkins does an interview wherein he describes a very satisfying explanation for how the eye evolved, likely using water molecules as a primitive lens. Very insightful.

Websites and Online Journalism

"A Groundbreaking Study Is Good News for Cats—And People". Yong, Ed. 07/10/2019. https://www.theatlantic.com/science/archive/2019/07/groundbreaking-parasite-study-good-news-cats/593779/

"Groundbreaking" is a little hyperbolic but nonetheless a fascinating article on why T. Gondii only reproduces in the cat as opposed to any other animal. Researchers were actually able to reproduce those conditions and make the parasite reproduce in mice.

"Americans Are Losing Faith in Democracy-and in Each Other." Persily, Nathaniel. Cohen, Jon. 10/14/16. https://www.washingtonpost.com/opinions/americans-are-losing-faith-in-democracy--and-in-each-other/2016/10/14/b35234ea-90c6-11e6-9c52-0b10449e33c4_story.html

"Americans Don't Trust Their Institutions Anymore." Malone, Claire. 11/16/16. https://fivethirtyeight.com/features/americans-dont-trust-their-institutions-anymore/

"Americans Have Lost Faith in Their Institutions. And That's Not Because of Trump or 'Fake News'." Bishop, Bill. 03/21/17. https://www.washingtonpost.com/posteverythin

g/wp/2017/03/03/americans-have-lost-faith-in-institutions-thats-not-because-of-trump-or-fake-news/

"Americans' Trust in Political Leaders, Public at New Lows." Jones, Jefferey. 09/01/16. https://news.gallup.com/poll/195716/americans-trust-political-leaders-public-new-lows.aspx

The Above four sources I will explain together since they are largely redundant and based off the same series of studies. Trust in institutions became a hot topic in the media around the time Trump became president and these media outlets wrote op-eds on the topic. I tried to get a broad selection and was pleased to see both sides of the political establishment give their opinions on the unsettling trend.

Atheists. Religious Landscape Study. https://www.pewforum.org/religious-landscape-study/religious-family/atheist/

Another fantastic study by Pew on religious affiliation and non-affiliation in the US.

The Discovery Institute. https://intelligentdesign.org/

> I don't fully agree with all of the points made therein, but a useful resource on the idea of intelligent design. I value their scholarship and arrogant refusal to quit in the face of so many dishonest enemies.

"Charles Colson, Nixon's 'dirty tricks' man, dies at 80." Dobbs, Michael. 04/20/2012. https://www.washingtonpost.com/politics/whitehouse/chuck-colson-nixons-dirty-tricks-man-dies-at-80/2012/04/21/gIQAaoOHYT_story.html

> The Washington Post Obituary on Charles Colson. I could not remember his name so this was the article I found in my google search. I later used it as a primer to his book, cited above.

"Christianity's Growth in China and Its Contributions to Freedoms." Yang, Fenggang. 10/31/2017. https://berkleycenter.georgetown.edu/responses/christianity-s-growth-in-china-and-its-contributions-to-freedoms

> A great non-biased piece from the very liberal Georgetown University on the growth of Christianity in China in spite of the totalitarian regime.

"Enlightenment."
> https://www.history.com/topics/british-history/enlightenment

> History.com reminds me of the golden age of the History channel before they sold their souls to the devil and became the channel of fat white men getting angry at each other on camera.

"From Genes to Species: A Primer on Evolution"
> Zivkovic, Bora. 09/24/2011.
> https://blogs.scientificamerican.com/a-blog-around-the-clock/bio101-from-genes-to-species-a-primer-on-evolution/

> Fantastic blow by blow account of the fundamentals of evolution.

"From Ideology to Racism: Hitler's Mein Kampf."
> https://web.nli.org.il/sites/nli/english/collection

s/personalsites/israel-germany/weimar-republic/pages/mein-kampf.aspx

Excellent article on Hitler's degeneration from his bigoted and bitter ideology to the monster that slaughtered millions.

"Gender Pronouns."

https://www1.nyc.gov/assets/hra/downloads/pdf/services/lgbtqi/Gender%20Pronouns%20final%20draft%2010.23.17.pdf

I was researching how far the Gender-pronoun argument had gone. It was in relationship to another rabbit hole I had gone down, even though this website doesn't make it into the final edition of the book.

"Jean-Jacques Rousseau."

https://www.britannica.com/biography/Jean-Jacques-Rousseau

Britannica is as good a reference as ever, and this citation went along with my research into the Enlightenment.

"Is Religion the Cause of Most Wars?" Rabbi Alan Laurie. 04/10/12. https://www.huffpost.com/entry/is-religion-the-cause-of-_b_1400766.

As I have said before this is likely the last time in my career that the Huffington Post will appear in my writing, but a solid Op-Ed nonetheless on the idea of religion and warfare.

"Researchers Develop Model of Toxoplasmosis Evolution." Packham, Christopher. 07/08/18. https://phys.org/news/2018-07-toxoplasmosis-evolution.html

I cite this to confirm a claim in my book that there is no real model for the evolution of T. Gondii. This model is woefully inaccurate and is lacking key information, or really any information, as to how a parasite began to control minds.

"Righteous Among the Nations." https://www.yadvashem.org/righteous/stories.html.

If you want to cry go through this list of saints who either risked or gave up their lives to save Jews in the Holocaust. Every American teenager should have to write a report about one of these heroes.

"Sam Harris, the New Atheist with a Spiritual Side" Anthony, Andrew. 02/16/19. https://www.theguardian.com/books/2019/feb/16/sam-harris-interview-new-atheism-four-horsemen-faith-science-religion-rationalism

It never ceases to amaze me that these interviews and podcasts with Harris can talk about spirituality, but he stays so married to Atheism. An Agnostic in denial if you ask me.

"Stephen Hawking Says He Knows What Happened Before the Big Bang". Spektor, Brandon. 03/02/2018. https://www-livescience-com.cdn.ampproject.org/v/s/www.livescience.com /amp/61914-stephen-hawking-neil-degrasse-tyson-beginning-of-time.html?amp_js_v=0.1&usqp=mq331AQCKAE%3D

Hawking has interesting thought experiments in this article but are obviously fallacious as I point out in the book.

"Scientists and Belief." 011/05/2009
> https://www.pewforum.org/2009/11/05/scientists-and-belief/
>
> Another great Pew study that very serendipitously examines belief in the sciences.

"Schilling Throws Wild Pitch with Nazi Stat." Greenburg, Jon. 08/25/15
> https://www.politifact.com/factchecks/2015/aug/28/curt-schilling/schilling-throws-wild-pitch-nazi-stat/
>
> A conservative uses the statistics about Hitler and Atheism wrongly. It caused a stir on Twitter with a few thousand people. That's the full story.

"Sex Reassignment Doesn't Work. Here Is the Evidence." Anderson, Ryan. 03/08/2018.

https://www.dailysignal.com/2018/03/08/sex-reassignment-doesnt-work-evidence/

Another citation for the rabbit hole I went down on the Trans movement. It ties in loosely with regards to the search for, and defining of, objective truth.

"The Parasite That Makes a Rat Love a Cat". Zielinski, Sarah. 09/22/2011.

https://www.smithsonianmag.com/science-nature/the-parasite-that-makes-a-rat-love-a-cat-86515093/

Smithsonian magazine is vastly underrated. Many great and well researched topics, such as this one on Toxoplasma Gondii, that I feel go widely unnoticed.

"The Bertrand Russell Archives."

https://www.mcmaster.ca/russdocs/russell.htm

I used this source to hunt down the proper quoting for the legendary Atheist thinker.

"Top Cosmologist's Lonely Battle Against 'Big Bang' Theory."
https://web.archive.org/web/20191114152141/https://www.afp.com/en/news/826/top-cosmologists-lonely-battle-against-big-bang-theory-doc-1m915e1

This is where I got the citation for the Nobel Prize winner who feels we don't know how the universe was started.

"Toxoplasmosis."
https://www.cdc.gov/parasites/toxoplasmosis/gen_info/faqs.html

The CDC page for the disease T. Gondii that causes in humans.

"Trust and Accuracy." Mitchell, Amy. Gottfried, Jefferey. Barthel, Michael. Shearer, Elisa. 07/01/16.
https://www.journalism.org/2016/07/07/trust-and-accuracy/

This study is widely cited for its accuracy and timeliness, especially in the tumultuous years around the 2016 American Presidential election.

"Trust Is Collapsing in America." Friedman, Uri. 01/21/18. https://www.theatlantic.com/international/archive/2018/01/trust-trump-america-world/550964/

This Op-Ed ran later than the first four in this section but is on a similar dismaying trend: the breakdown of dialogue between American citizens.

"U.S. death rates from suicides, alcohol and drug overdoses reach all-time high". Edwards, Erika. 06/12/2019. https://www.nbcnews.com/health/health-news/u-s-death-rates-suicides-alcohol-drug-overdoses-reach-all-n1016216

Great article on the topic of "Deaths of Despair" and the physical realities of Nihilism.

"What Is Time? One Physicist Hunts for the Ultimate Theory" Biba, Erin. https://www.wired.com/2010/02/what-is-time/

> Great article on the theoretical study of time and the pursuit of a hardened and testable theory. Also this interview of Sean Carroll led me to the book that this article is promoting, cited above.

"Why Everything You've Been Told About Evolution is Wrong". Burkeman, Oliver. 03/19/2010. https://www.google.com/amp/s/amp.theguardian.com/science 2010/mar/19/evolution-darwin-natural-selection-genes-wrong

> This article is one of many on the trend of fighting against the outdated idea of natural selection, although still it is an Atheist author. To some in the community there is increasing evidence that suggests a revised model for evolution and speciation is required.

"Three Decades Ago, America Lost Its Religion. Why?" Thompson, Derek. 09/26/19 https://www.theatlantic.com/ideas/archive/2019/09/atheism-fastest-growing-religion-us/598843/

A click-bait title but an overall good source on the down trend in casual church attendance and loose religious affiliation.

"The Myth of "Mind-Altering Parasite" Toxoplasma Gondii?." 02/20/2016. https://www.discovermagazine.com/health/the-myth-of-mind-altering-parasite-toxoplasma-gondii

A rebuttal to the notion that T. Gondii displays mind-altering affects in humans. Not a huge surprise as in my lightly informed opinion the previous studies on the topic were tenuous at best.

"Toxoplasmosis." https://www.avma.org/resources/pet-owners/petcare/toxoplasmosis

A primer on Toxoplasmosis in animals from the American Veterinary Medical Association.

"Toxoplasmosis: How a Cat Parasite Exploits Immune Cells to Reach the Brain." ScienceDaily. www.sciencedaily.com/releases/2017/12/171208095923.htm

> Interesting article on how T. Gondii hijacks immune cells to get past the body's defenses. Reads like it belongs in the "Scholarly Article" section unfortunately.

"Toxoplasmosis in Cats."
https://www.vet.cornell.edu/departments-centers-and-institutes/cornell-feline-health-center/health-information/feline-health-topics/toxoplasmosis-cats

> Ironically I had a hard time finding a primer on the effects of T. Gondii in cats. This was a good one by Cornell.

"7 Strange Facts about the 'Mind-Control' Parasite Toxoplasma Gondii." Bucklin, Stephanie. 10/18/16. https://www.livescience.com/56529-strange-facts-about-toxoplasma-gondii-parasite.html

I have become a big fan of live science through this process. They produce a lot of quality content, digestible for the lay public.

"10 Facts About Atheists" Lipka, Michael. 12/06/2019. https://www.pewresearch.org/fact-tank/2019/12/06/10-facts-about-atheists/

Further analysis on data about Atheist by Pew. I have been very impressed with their organization and lack of obvious bias in my research.

"100 Years of Communism—and 100 Million Dead." Adam Satter. 11/06/17. https://www.wsj.com/articles/100-years-of-communismand-100-million-dead-1510011810.

Great article on the near apocalyptic impact of Communism on its own people.

Scholarly and Peer-Reviewed Journals, Papers, Etc.

"Toxoplasma Gondii" Obstetrics & Gynecology: March 1963 - Volume 21 - Issue 3 - p 318-329

Fascinating early look at the research on *T. Gondii* back in the 60's, and through the lens of female health.

Bindu Gajria, Amit Bahl, John Brestelli, Jennifer Dommer, Steve Fischer, Xin Gao, Mark Heiges, John Iodice, Jessica C. Kissinger, Aaron J. Mackey, Deborah F. Pinney, David S. Roos, Christian J. Stoeckert, Jr, Haiming Wang, Brian P. Brunk, ToxoDB: an integrated Toxoplasma gondii database resource , Nucleic Acids Research, Volume 36, Issue suppl_1, 1 January 2008, Pages D553–D556, https://doi.org/10.1093/nar/gkm981

Dass, Shantala. Yves, Ajai. "Toxoplasma Gondii Infection Reduces Predator Aversion in Rats Through Epigenetic Modulation in the Host Medial Amygdala." Mol Ecol, 23: 6114-6122. doi:10.1111/mec.12888.

Daniel Ajzenberg, Nadine Cogné, Luc Paris, Marie-Hélène Bessières, Philippe Thulliez, Denis Filisetti, Hervé Pelloux, Pierre Marty, Marie-Laure Dardé, "Genotype of 86 Toxoplasma Gondii Isolates Associated with Human Congenital Toxoplasmosis, and Correlation with Clinical Findings". The Journal of Infectious Diseases, Volume 186, Issue 5, 1 September 2002, Pages 684–689, https://doi.org/10.1086/342663

Hill, J.P. Dubey, "Toxoplasma Gondii: Transmission, Diagnosis and Prevention". Clinical Microbiology and Infection, Volume 8, Issue 10, 2002, Pages 634-640, ISSN 1198-743X, https://doi.org/10.1046/j.1469-0691.2002.00485.x.

Ildiko Rita Dunay, Kiran Gajurel, Reshika Dhakal, Oliver Liesenfeld, Jose G. Montoya, "Treatment of Toxoplasmosis: Historical Perspective, Animal Models, and Current Clinical Practice". Clinical Microbiology Reviews Sep 2018, 31 (4) e00057-17; DOI: 10.1128/CMR.00057-17

Kim, Kami. Weiss, Louis. "Toxoplasma Gondii: The Model Apicomplexan. Perspectives and Methods". (Elsevier, 2007)

Kissinger, J. C., Gajria, B., Li, L., Paulsen, I. T., & Roos, D. S. (2003). "ToxoDB: accessing the Toxoplasma Gondii Genome." Nucleic Acids Research, 31(1), 234–236. doi:10.1093/nar/gkg072

Liu, Q., Wang, Z., Huang, S. et al. "Diagnosis of Toxoplasmosis and Typing of Toxoplasma Gondii". Parasites Vectors 8, 292 (2015). https://doi.org/10.1186/s13071-015-0902-6

Mineo, J.R., Kasper, L.H. "Attachment of Toxoplasma gondii to Host Cells Involves Major Surface

Protein, SAG-1 (P-30)." Experimental Parasitology, Volume 79, Issue 1, August 1994, Pages 11-20

Miró G, Montoya A, Jiménez S, Frisuelos C, Mateo M, Fuentes I. Prevalence of Antibodies to Toxoplasma Gondii and Intestinal Parasites in Stray, Farm and Household Cats in Spain. Vet Parasitol. 2004;126(3):249-255. doi:10.1016/j.vetpar.2004.08.015

Rahimi, M. T., Daryani, A., Sarvi, S., Shokri, A., Ahmadpour, E., Teshnizi, S. H., Mizani, A., & Sharif, M. (2015). "Cats and Toxoplasma Gondii: A Systematic Review and Meta-Analysis in Iran". The Onderstepoort journal of veterinary research, 82(1), e1–e10. https://doi.org/10.4102/ojvr.v82i1.823

Silva, Rodrigo Costa da, Langoni, Helio, & Megid, Jane. "Adaptive and Genetic Evolution of Toxoplasma Gondii: A Host-Parasite Interaction" 2007. Revista da Sociedade

Brasileira de Medicina Tropical, 50(4), 580-581. https://dx.doi.org/10.1590/0037-8682-0251-2017

Solène Rougier, Jose G. Montoya, François Peyron, "Lifelong Persistence of Toxoplasma Cysts: A Questionable Dogma?", Trends in Parasitology, Volume 33, Issue 2, 2017, Pages 93-101, ISSN 1471-4922, https://doi.org/10.1016/j.pt.2016.10.007.

Stock, A., Dajkic, D., Köhling, H. et al. Humans With Latent Toxoplasmosis Display Altered Reward Modulation of Cognitive Control". Sci Rep 7, 10170 (2017). https://doi.org/10.1038/s41598-017-10926-6

Yanhua Wang, Guangxiang Wang, Delin Zhang, Hong Yin, Meng Wang, "Identification of Novel B Cell Epitopes Within Toxoplasma Gondii GRA1." Experimental Parasitology, Volume 135, Issue 3, 2013, Pages 606-610, ISSN 0014-4894, https://doi.org/10.1016/j.exppara.2013.09.019.

Yaw Adomako-Ankomah, Elizabeth D. English, Jeffrey J. Danielson, Lena F. Pernas, Michelle L. Parker, Martin J. Boulanger, Jitender P. Dubey and Jon P. Boyle, "Host Mitochondrial Association Evolved in the Human Parasite Toxoplasma gondii via Neofunctionalization of a Gene Duplicate". Genetics May 1, 2016 vol. 203 no. 1 283-298; https://doi.org/10.1534/genetics.115.186270

Ze-Dong Wang, Shu-Chao Wang, Huan-Huan Liu, Hong-Yu Ma, Zhong-Yu Li, Feng Wei, Xing-Quan Zhu, Quan Liu, "Prevalence and Burden of Toxoplasma Gondii Infection in HIV-Infected People: a Systematic Review and Meta-Analysis". The Lancet HIV, Volume 4, Issue 4, 2017, Pages e177-e188, ISSN 2352-3018, https://doi.org/10.1016/S2352-3018(17)30005-X.

For the sources on *T. Gondii* I will summarize them here because they are all obviously on the same general topic, not very long articles, and most importantly they tell you in the title what they're about. They're all excellent (as far the hell as I know) if fairly dry. I understand that scholarly papers are not prose, but I don't think it would kill anyone to put some thought into the flow of communication.

Quattrociocchi, Walter and Scala, Antonio and Sunstein, Cass R., "Echo Chambers on Facebook". June 13, 2016. Available at SSRN: https://ssrn.com/abstract=2795110 or http://dx.doi.org/10.2139/ssrn.2795110

Excellent scientific look at what social media is doing to us. Spoiler: it isn't good.

Schnabel, Landon, and Sean Bock. 2017. "The Persistent and Exceptional Intensity of American Religion: A Response to Recent Research." Sociological Science 4:686-700.

A solid counter to the idea religion is waning in America. According to this article it is actually just the casually religious who are sliding into the "none" category, whereas before they were just nominally religious.

"Vertebrae Flight."

https://ucmp.berkeley.edu/vertebrates/flight/evolve.html

A great piece by UCMP Berkeley talking about the theoretical origins of flight. Obviously it does not answer that question without a mention of a creator, but interesting nonetheless.

Connect with Matt

Well here we are folks, the back of the book. If you found your way to the end and are looking for more there are still a couple ways to scratch that itch before I add to the little shelf in my office I bought to hold all the books I would write. First, you can check out my annotated bibliography for a few books I highly recommend. If you haven't read *Mere Christianity* or *The Screwtape Letters* then I'm sorry you chose this first. They're old books, you can find them at any library, and C.S. Lewis is likely one of the greatest Christian Apologists ever.

Next, you can follow me on Instagram for pictures of my adventures and the wildly (surprisingly) popular *Tuesday Night Hottakes* (I can't explain it, just follow me and find out). For some political thoughts and witty (in my opinion) one-liners, follow me on Twitter. My *Medium* account is behind the pay-wall (I know, I'm sorry, we all gotta eat) but I treat it like an opinion section of a major news publication and hold myself to high standards (in that case, maybe that's not the right example). Most of the time. All accounts have the same username: @mattaloveland.

$10 bucks is no joke sometimes, if you know someone (a younglife girl, your second cousin, the mailman, etc.) who needs this book but doesn't have the money, just let me know. I wrote this book in faith, God won't let me starve.

Finally, and I have been kicking this one around for awhile, if you'd like to *literally* connect with me, here's my phone number: (803) 257-7517. Yep. That's it. One of my favorite authors, Bob Goff, put it in his book, *Love Does*, and if that *New York Times* Bestselling author can be brave, then so can I. Let me know if there's anything I can help with.

About the Author

Matt Loveland is a graduate of The Citadel, the Military College of South Carolina, and serves in the National Guard. He began writing after working for *Younglife* in Charleston, SC. His time coaching and leading young men at local high schools led him to question the problems at the heart of our society. This book is the result of that journey.

www.ingramcontent.com/pod-product-compliance
Lightning Source LLC
Chambersburg PA
CBHW060150050426
42446CB00013B/2748